Berdimurodov Elyor Tukhlievich and Chandrabhan Verma (Eds.)

Corrosion Prevention Nanoscience

Also of interest

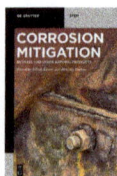

Corrosion Mitigation.
Biomass and Other Natural Products
Ashish Kumar and Abhinay Thakur (Eds.), 2022
ISBN 978-3-11-076057-6, e-ISBN 978-3-11-076058-3

Applied Electrochemistry
Krystyna Jackowska and Paweł Krysiński, 2020
ISBN 978-3-11-060077-3, e-ISBN 978-3-11-060083-4

Electrochemistry.
A Guide for Newcomers
Helmut Baumgärtel, 2019
ISBN 978-3-11-044340-0, e-ISBN 978-3-11-043739-3

Materials Corrosion and Protection
Yongchang Huang and Jianqi Zhang (Eds.), 2018
ISBN 978-3-11-030987-4, e-ISBN 978-3-11-038295-2

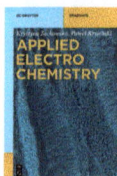

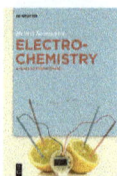

Corrosion Prevention Nanoscience

Nanoengineering Materials and Technologies

Edited by
Berdimurodov Elyor Tukhlievich and Chandrabhan Verma

DE GRUYTER

Editors
Berdimurodov Elyor Tukhlievich
Faculty of Chemistry
National University of Uzbekistan
Tashkent 100034
Usbekistan
Email: elyor170690@gmail.com

Dr. Chandrabhan Verma
Department of Chemical Engineering
Khalifa University of Science and Technology
P.O. Box 127788
Abu Dhabi
United Arab Emirates
Email: Chandraverma.rs.apc@itbhu.ac.in

ISBN 978-3-11-107009-4
e-ISBN (PDF) 978-3-11-107175-6
e-ISBN (EPUB) 978-3-11-107189-3

Library of Congress Control Number: 2023935911

Bibliographic information published by the Deutsche Nationalbibliothek
The Deutsche Nationalbibliothek lists this publication in the Deutsche Nationalbibliografie;
detailed bibliographic data are available on the internet at http://dnb.dnb.de.

© 2023 Walter de Gruyter GmbH, Berlin/Boston
Cover image: Andriy Onufriyenko/Moment/Getty Images
Typesetting: Integra Software Services Pvt. Ltd.
Printing and binding: CPI books GmbH, Leck

www.degruyter.com

Contents

List of authors

Nurşah Kütük
Chemical Engineering Department
Faculty of Engineering
Sivas Cumhuriyet University
Sivas
Turkey

Aslıhan Gürbüzer
Plant and Animal Production Department
Technical Sciences Vocational School of Sivas
Sivas Cumhuriyet University
Sivas
Turkey

Gamze Tüzün
Department of Chemistry
Faculty of Science
Sivas Cumhuriyet University
Sivas
Turkey
Email: gamzekekul@gmail.com

Burak Tüzün
Plant and Animal Production Department
Technical Sciences Vocational School of Sivas
Sivas Cumhuriyet University
Sivas
Turkey

Omar Dagdag
Centre for Materials Science
College of Science
Engineering and Technology
University of South Africa
Johannesburg 1710
South Africa

Rajesh Haldhar
School of Chemical Engineering
Yeungnam University
Gyeongsan 38541
Republic of Korea

Seong-Cheol Kim
School of Chemical Engineering
Yeungnam University
Gyeongsan 38541
Republic of Korea

Walid Daoudi
Laboratory of Molecular Chemistry
Materials and Environment (LCM2E)
Department of Chemistry
Multidisciplinary Faculty of Nador
University Mohamed I
60700 Nador
Morocco

Elyor Berdimurodov
Faculty of Chemistry
National University of Uzbekistan
Tashkent 100034
Uzbekistan
Email: elyor170690@gmail.com

Ekemini D. Akpan
Centre for Materials Science
College of Science
Engineering and Technology
University of South Africa
Johannesburg 1710
South Africa

Eno E. Ebenso
Centre for Materials Science
College of Science
Engineering and Technology
University of South Africa
Johannesburg 1710
South Africa

Khasan Berdimuradov
Faculty of Industrial Viticulture and Food
Production Technology
Shahrisabz branch of Tashkent Institute of
Chemical Technology
Shahrisabz 181306
Uzbekistan

https://doi.org/10.1515/9783111071756-203

Ilyos Eliboev
Faculty of Chemistry
National University of Uzbekistan
Tashkent 100034
Uzbekistan

Lazizbek Azimov
Faculty of Chemistry
National University of Uzbekistan
Tashkent 100034
Uzbekistan

Yusufboy Rajabov
Faculty of Chemistry
National University of Uzbekistan
Tashkent 100034
Uzbekistan

Jaykhun Mamato
Faculty of Chemistry
National University of Uzbekistan
Tashkent 100034
Uzbekistan

Bakhtiyor Borikhonov
Faculty of Chemistry–Biology
Karshi State University
Karshi 130100
Uzbekistan

Abduvali Kholikov
Faculty of Chemistry
National University of Uzbekistan
Tashkent 100034
Uzbekistan

Khamdam Akbarov
Faculty of Chemistry
National University of Uzbekistan
Tashkent 100034
Uzbekistan

Aysun Aksu
Department of Molecular Biology and Genetics
Science Faculty
Sivas Cumhuriyet University
58140 Sivas
Turkey

Hüseyin Fatih Çetinkaya
Department of Environmental Engineering
Faculty of Engineering
Sivas Cumhuriyet University
Sivas
Turkey

Serap Çetinkaya
Department of Molecular Biology and Genetics
Science Faculty
Sivas Cumhuriyet University
58140 Sivas
Turkey
Email: serapcetinkaya2012@gmail.com

Gamze Tüzün
Department of Chemistry
Faculty of Science
Sivas Cumhuriyet University
Sivas
Turkey

Burak Tüzün
Plant and Animal Production Department
Technical Sciences Vocational School of Sivas
Sivas Cumhuriyet University
Sivas
Turkey

Bo-kai Liao
School of Chemistry and Chemical Engineering
Guangzhou University
Guangzhou 510006
China
and
Joint Institute of Guangzhou University &
Institute of Corrosion Science and Technology
Guangzhou University
Guangzhou 510006
China
Email: bokailiao@gzhu.edu.cn

Zhi-Gang Luo
School of Chemistry and Chemical Engineering
Guangzhou University
Guangzhou 510006
China

Shan Wan
School of Chemistry and Chemical Engineering
Guangzhou University
Guangzhou 510006
China
and
Joint Institute of Guangzhou University &
Institute of Corrosion Science and Technology
Guangzhou University
Guangzhou 510006
China

Hao-Wei Deng
School of Physics and Materials Science
Center of Advanced Functional Materials
Guangzhou University
Guangzhou 510006
China

Shu-Yi Jiang
School of Art & Design
Guangdong University of Technology
Guangzhou 510062
China

Shuang-Jian Li
Institute of New Materials
Guangdong Academy of Sciences
National Engineering Laboratory of Modern
Materials Surface Engineering Technology
Guangzhou 510650
China

Jun-Jie Yang
Institute of Advanced Wear and Corrosion
Resistant and Functional Materials
Jinan University
Guangzhou 510632
China

Humira Assad
Department of Chemistry
School of Chemical Engineering and Physical
Sciences
Lovely Professional University
Punjab
India

Ashish Kumar
NCE
Department of Science and Technology
Bihar Engineering University
Government of Science and Technology
Government of Bihar
Bihar
India

Dheeraj Singh Chauhan
Modern National Chemicals
Second Industrial City
Dammam 31421
Saudi Arabia
Email: dheeraj.chauhan.rs.apc@itbhu.ac.in

Dakeshwar kumar Verma
Department of Chemistry
Government Digvijay Autonomous Postgraduate
College
Rajnandgaon 491441
Chhattisgarh
India

Reema Sahu
Department of Chemistry
Government Digvijay Autonomous Postgraduate
College
Rajnandgaon 491441
Chhattisgarh
India

Santosh Bahadur singh
Department of Chemistry
University of Allahabad
Prayagraj 211002
Uttar Pradesh
India

Bharti Yarda
Department of Chemistry
Government Digvijay Autonomous Postgraduate
College
Rajnandgaon 491441
Chhattisgarh
India

Vikas Kumar Jain
Department of Technical Education Indravati
Bhawan
Nava Raipur
Atal Nagar
Raipur 492002
Chhattisgarh
India

Shailendra Yadav
Department of Chemistry
AKS University
Satna
Madhyapradesh
India
and
Department of Chemistry
Gurughasidas Central University
Bilaspur
Chhattisgarh
India

Vikash Kande
Department of Chemistry
Government Digvijay Autonomous Postgraduate
College
Rajnandgaon 491441
Chhattisgarh
India

Sharad Tiwari
Department of Chemistry
Government Digvijay Autonomous Postgraduate
College
Rajnandgaon 491441
Chhattisgarh
India

Gokul Ram Nishad
Department of Chemistry
Government Digvijay Autonomous Postgraduate
College
Rajnandgaon 491441
Chhattisgarh
India

Younus Raza Beg
Department of Chemistry
Government Digvijay Autonomous Postgraduate
College
Rajnandgaon 491441
Chhattisgarh
India

Vandana Mishra
Department of Chemistry
Government Digvijay Autonomous Postgraduate
College
Rajnandgaon 491441
Chhattisgarh
India

Durgesh Sinha
Department of Chemistry
Gurughasidas Central University
Bilaspur
Chhattisgarh
India

Muhammed Safa Çelik
Department of Molecular Biology and Genetics
Science Faculty
Sivas Cumhuriyet University
58140 Sivas
Turkey

Hüseyin Fatih Çetinkaya
Department of Environmental Engineering
Faculty of Engineering
Sivas Cumhuriyet University
Sivas
Turkey

Serap Çetinkaya
Department of Molecular Biology and Genetics
Science Faculty
Sivas Cumhuriyet University
58140 Sivas
Turkey

Gamze Tüzün
Department of Chemistry
Faculty of Science
Sivas Cumhuriyet University
Sivas
Turkey

Burak Tüzün
Plant and Animal Production Department
Technical Sciences Vocational School of Sivas
Sivas Cumhuriyet University
Sivas
Turkey

Amir Hossein Jafari Mofidabadi
Department of Chemical Engineering
Faculty of Engineering
Golestan University
Aliabad Katoul
Iran

Nariman Alipanah
Department of Surface Coatings and Corrosion
Institute for Color Science and Technology (ICST)
P.O. 16765-654
Tehran
Iran

Ali Dehghani
Department of Chemical Engineering
Faculty of Engineering
Golestan University
Aliabad Katoul
Iran
Email: dehghaniali1996@gmail.com

Manash Protim Mudoi
Department of Chemical Engineering
Indian Institute of Technology
Roorkee 247667
Uttarakhand
India
and
Department of Chemical Engineering
University of Petroleum and Energy Studies
Dehradun 248007
Uttarakhand, India
Email: mp_mudoi@ch.iitr.ac.in, mpmudoi@ddn.
upes.ac.in

Rhythm Katyal
Department of Chemical Engineering
University of Petroleum and Energy Studies
Dehradun 248007
Uttarakhand
India

Khushi Bhatt
Department of Chemical Engineering
University of Petroleum and Energy Studies
Dehradun 248007
Uttarakhand
India

Vidushi Singh
Department of Chemical Engineering
University of Petroleum and Energy Studies
Dehradun 248007
Uttarakhand
India

Asmita Choudhary
Department of Chemical Engineering
University of Petroleum and Energy Studies
Dehradun 248007
Uttarakhand
India

Sanskriti Gupta
Department of Chemical Engineering
University of Petroleum and Energy Studies
Dehradun 248007
Uttarakhand
India

Khasan Berdimuradov
Faculty of Industrial Viticulture and Food
Production Technology
Shahrisabz branch of Tashkent Institute of
Chemical Technology
Shahrisabz 181306
Uzbekistan

Ilyos Eliboev
Faculty of Chemistry
National University of Uzbekistan
Tashkent 100034
Uzbekistan

Nurbek Umirov
Faculty of Chemistry–Biology
Karshi State University
Karshi 130100
Uzbekistan

Bakhtiyor Borikhonov
Faculty of Chemistry–Biology
Karshi State University
Karshi 130100
Uzbekistan

Abduvali Kholikov
Faculty of Chemistry
National University of Uzbekistan
Tashkent 100034
Uzbekistan

Khamdam Akbarov
Faculty of Chemistry
National University of Uzbekistan
Tashkent 100034
Uzbekistan

Nurşah Kütük, Aslıhan Gürbüzer, Gamze Tüzün*, Burak Tüzün

Chapter 1
Nanoengineering and nanoscience: current and emerging trends

Abstract: The study of science at the nanoscale has become one of the most important research topics in recent years. Almost all branches of science are interested in nanotechnology. Metal and metal oxides, carbon family products, polymers, clays and composites are some of the nanomaterials used. Despite their small size, nanomaterials have different properties such as large surface area, different particle shape and size, pore volume, biocompatibility and biodispersibility. For this reason, nanotechnology has also spread to the industry. Due to their unique properties, nanomaterials are widely used in medical, water treatment, sensor and environmental applications. Although nanotechnology research has accelerated especially in recent years, there are still many unexplored materials and application areas. The aim of this study is to investigate the advantages and disadvantages of nanomaterials and nanotechnology applications by examining their usage areas in scientific and industrial fields.

Keywords: Nanoparticles, nanotechnology, materials, metal oxides, particle size

1.1 Introduction

The study of nanoscale materials and the development of associated laws and conceptual information constitute the field of science known as nanoscience. Generally, materials smaller than 100 nm are classified as nanomaterials [1]. Its scientific equivalent is a billionth of a meter [2]. In fact, the existence of nanoparticles (NPs) is not a new phenomenon. Nanoscale materials have been formed in nature for centuries. Some of these are the presence of events that are sources of NPs such as volcanic eruptions and fires and ceramic glazes or stained glass silver NPs for color purposes [3]. The scientific development of NPs for the first time was around 50 years ago for vaccines and cancer treatment [4]. Nanotechnology cannot be limited to any discipline. It is interdisciplinary with natural sciences and various branches of engineering and sciences such as toxicology

*Corresponding author: Gamze Tüzün, Department of Chemistry, Faculty of Science, Sivas Cumhuriyet University, Sivas, Turkey, e-mail: gamzekekul@gmail.com, http://orcid.org/0000-0002-0420-2043
Nurşah Kütük, Chemical Engineering Department, Faculty of Engineering, Sivas Cumhuriyet University, Sivas, Turkey
Aslıhan Gürbüzer, Burak Tüzün, Plant and Animal Production Department, Technical Sciences Vocational School of Sivas, Sivas Cumhuriyet University, Sivas, Turkey

https://doi.org/10.1515/9783111071756-001

[3]. In recent years, nanotechnology has been used in different disciplines such as engineering, physics, chemistry and medicine. It draws attention with its diagnostic, imaging or therapeutic properties, especially in the biomedical field [5, 6].

The structural properties of nanoparticles (NPs) and the area to be used vary depending on their properties such as size, shape, crystallinity and morphology [7]. An important feature that makes NPs stand out is that they have a larger surface area compared to the raw materials from which they are synthesized. Using many organic and inorganic elements, NPs can be synthesized from natural components such as lipids, proteins, polysaccharides or metals such as Ti, Ag, Au, Zn and Cu by various methods [8, 9]. The use of NPs in the diagnosis and treatment of diseases such as cancer reveals the importance of nanotechnology due to their biocompatibility and biodispersibility properties [7]. As seen in Figure 1.1, a glucose molecule is 1 nm in size, while a water molecule is 0.1 nm. These data indicate that nanomaterials are at the molecular size level [10].

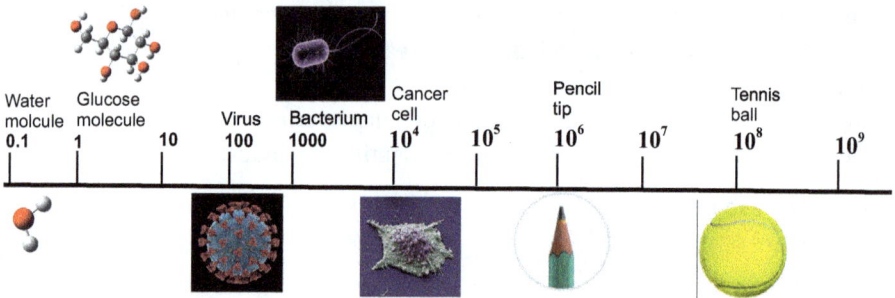

Figure 1.1: Comparison of length scale 10^9.

1.2 Nanomaterials

Nanomaterials with a high surface/volume ratio form the basis of nanotechnology due to their many superior properties. Researches have reported that there are over 1,000 nanoproducts due to their wide usage areas [11]. Today, many material groups such as NPs, nanotubes, nanomembranes and nanofibers are investigated [12]. Numerous distinct metallic NPs have recently been developed and applied. Titanium dioxide (TiO_2), zinc oxide (ZnO), FeO, copper oxide (CuO), bismuth oxide (Bi_2O_3), silica, mesoporous silica and lead oxide (PbO) are some of them [13–15]. Zirconium dioxide (ZrO_2), which is one of the important industrial nanomaterials, has important properties such as catalytic activity, thermal stability and chemically inert structure. Cobalt oxide (Co_3O_4) is another important material in memory storage units, catalytic activity and chemical sensor applications [16]. Zinc NPs are an important semiconductor with optical, piezoelectric and nontoxic properties [7, 17]. Silica and mesoporous silica NPs are nontoxic, biocompatible and biodegradable materials. Thanks to its large surface area and pore volume, it can be used in different applications such as drug release, bone tissue engineering, adsorption, biosensor and solar cell [15]. Various metallic, carbon, polymeric or composite NPs used in recent years and their application areas are given in Table 1.1.

Carbon is a rich and important family for nanoscale systems and exhibits broad functionalities with structural arrangements [18]. sp^2 carbon materials can generally be thought of as zero-dimensional fullerene, one-dimensional carbon nanotube (CNT) and two-dimensional graphene [19]. Graphene has unique electronic, optical, catalytic, mechanical and magnetic properties, thermal conductivity and gas barrier properties. Thanks to these properties, graphene has uses in different applications from energy storage to biomedical materials [20, 21]. Graphene oxide (GO) and reduced graphene oxide (RGO) are known as several-layer graphene oxide materials [22]. In addition, carbon-based nanomaterials such as fullerene and CNT are used in different applications from medicine to electronics such as metal NPs [23]. Figure 1.2 shows the molecular structure and shape of fullerene, CNT and graphene.

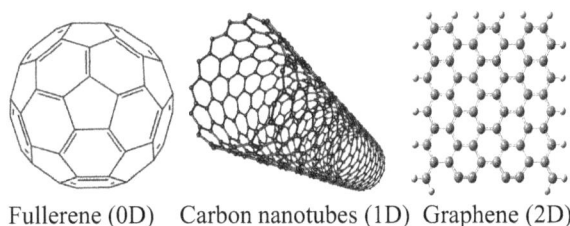

Fullerene (0D)   Carbon nanotubes (1D) Graphene (2D)

Figure 1.2: Moleculer structure of fullerene, carbon nanotubes and graphene.

Table 1.1: Different nanoparticles and area of applications.

Nanoparticles	Area of application	References
TiO_2	Bone tissue engineering	[24]
TiO_2	Antibacterial, larvicidal and anticancer effects	[8]
CeO_2	Anticancer effect	[25]
ZnO	Antifungal activity	[17]
Chitosan nanoparticles	Antibacterial activity	[26]
Silane-modified Fe_2O_3	Anticorrosion	[27]
CuO NPs	Adsorption	[28]
NiO NPs	Adsorption	[28]
Silver	Catalytic activity	[29]
CuO NPs	Photocatalytic, antimicrobial and anticancer activity	[30]
Fe/Zn bimetallic	Cytotoxicity study	[7]
Mesoporous-ordered silica (MCM-41)	Adsorption	[31]
$ZnO/ZnCr_2O_4$	Photocatalytic degradation	[32]
Graphene	Cytocompatibility effect	[20]
$ZrO_2/Co_3O_4/RGO$	Electrochemical sensor	[16]
PEGlyted nanographene	Photothermal therapy Targeted chemotherapy	[21]

In addition, nano-sized materials of clays have come to the fore in recent years. Nanoclays can be used in drug carrier systems, catalyst or adsorbent, food packaging in different areas. The nanosize of the clay affects the thermodynamic, physical and chemical properties of the composites synthesized with polymers. It provides benefits such as increasing the stability of the drug or increasing the biodegradability of the polymer in drug release applications [33]. Polymer nanoparticles can be obtained both from natural polymers and synthetically by synthesis. They can be functionalized with polysaccharide, protein or amino acid molecules. In this way, they become more effective for drug release [34]. Polymeric nanoparticles with high drug loading capacity can be easily sterilized and dispersed [35]. Poly-hydroxyethyl methacrylate (p(HEMA)), which can be synthesized in nanosize, is an important polymer in this field and is used in biomedical and pharmaceutical fields such as soft contact lenses [34]. Chitosan NPs, on the other hand, have high zeta potential and large surface area. The use of chitosan nanoparticles, which are used in the treatment of agricultural diseases, in many areas such as drugs, vaccines and gene release is being investigated [26].

Nanofibers are very fine yarns with a diameter between 50 and 500 nm, which are light, porous and have a high activity. They are unique materials for dressings, protective clothing, membrane and tissue engineering. They are obtained by different methods such as phase separation, interfacial polymerization, freeze-drying synthesis, especially electrospinning method [36]. In addition, the production of nano-based cellulose fibers and their applications in composite materials has increased interest due to important properties such as high stiffness, biodegradability and low weight. The application of cellulose nanofibers in polymer reinforcement is a topic of recent interest [37].

Nanocomposites with organic or inorganic properties are functional materials that do not form a mixture with each other and contain organic and inorganic components in their structures. Complex nanometer-based structures can be formed from them [38].

1.2.1 Synthesis methods and structure analyses

There are two different approaches to synthesis methods in nanotechnology. One of them is the "top-down" approach. In this approach, nano-based materials are made using the smallest structures. This is photonic applications in nanoelectricity and nanoengineering. The "bottom-up" approach, on the other hand, can be called molecular nanotechnology. Science is concerned with the production of different nanoparticles and nanomaterials with extraordinary properties [39].

With the increase in research on the subject, the synthesis methods of nanoparticles have also been developed by scientists. Green synthesis, which has become popular in recent years, is frequently preferred because it is environmentally friendly, inexpensive, efficient and easy. Plant, fruit and vegetable extracts, bacteria and various microorganisms are used as reducing agents. Various phytochemical substances such as polyphenols, especially in plant extracts, reduce metal ions and provide nanoparticle synthesis [9]. Some of the commonly used methods are given in Figure 1.3.

After synthesis, the chemical structure of nanoparticles can be analyzed by techniques such as
- ultraviolet spectroscopy (UV/vis),
- Fourier transform spectroscopy,
- x-ray diffraction,
- energy-dispersive X-ray spectroscopy.

Properties such as size and particle shape in the morphological structure can be determined using
- scanning electron microscope,
- transmmison electron microscope (TEM),
- atomic force microscope,
- selected field electron diffraction techniques [9, 32, 40].

TEM images are given for superparamagnetic Fe_3O_4 NPs in Figure 1.4a and for hydroxyapatite NPs in Figure 1.4b. When the morphological structure of nanoparticles is examined, it is seen that the shape of Fe_3O_4 is spherical and hydroxyapatite is needle-like [41].

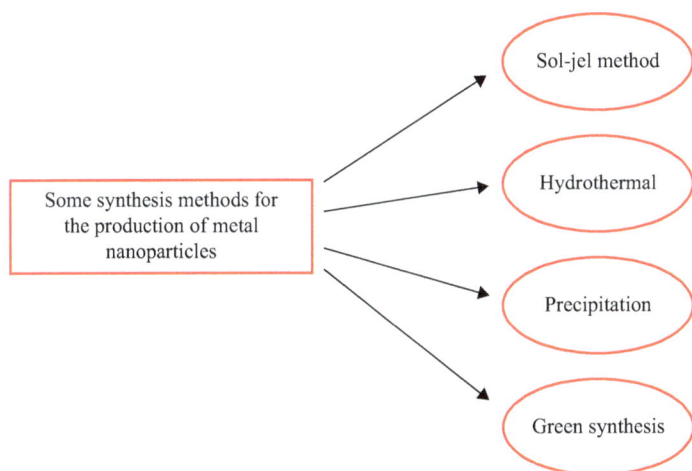

Figure 1.3: Some of the common methods used in metal nanoparticle production in recent years.

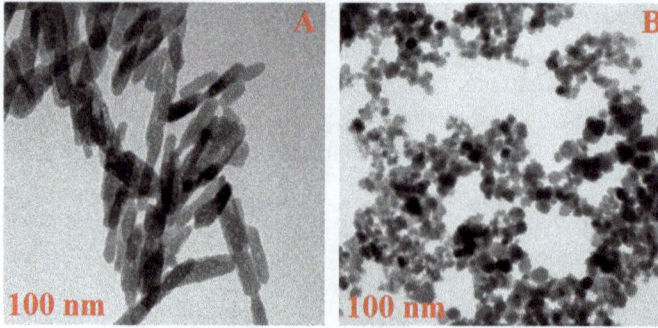

Figure 1.4: (a) Superparamagnetic Fe_3O_4 nanoparticles and (b) hydroxyapatite.

1.3 Scope of application

Nanoparticles obtained in organic, inorganic, metal or hybrid form have very good electronic, optical, physical or biological properties. Therefore, the use of nanoparticles has spread to various fields such as [3, 30, 39, 41]

– electronics,
– medicine (controlled drug delivery, anticancer activity, tissue repair, hemolytic activity),
– biomedical (pharmacy, detecing pathogen, antigen diagnosis),
– catalysis,
– fluid,
– agriculture.

In this section, popular applications and research areas in which nanomaterials are used today will be examined.

1.3.1 Medicine, health and biomedical areas

The use of nanotechnology in medicine has paved the way for the use of nanomaterials for the diagnosis of diseases or for various therapeutic purposes [42]. For example, cancer is a disease that has been affecting human beings for many years and whose treatment has been investigated. Cancer is the uncontrolled division and proliferation of cells. Sometimes it can occur in a certain organ and sometimes in a spread [43]. Especially for cancer treatment, many nanomaterials such as Ag NPs and TiO_2 NPs are used as drug delivery systems or for their therapeutic effects [8, 42].

The use of nanoparticles in bone tissue engineering has increased in recent years. Superparamagnetic iron oxide and hydroxyapatite nanoparticles were observed to improve mineralization and osteogenic differentiation of bone tissue [41]. Superparamagnetic iron oxide nanoparticles are used against brain tumor. It is shown in Figure 1.5 that the composite material synthesized in small nanoscale, modified with silica nanoparticles and CNTs passes through the blood-brain barrier (BBB). Even in a low magnetic field, nanoparticles have been reported to provide heat to destroy tumors. It has been reported that superparamagnetic iron oxide/silica/carbon nanoparticles (earthicles) pass through the BBB from the blood to the brain and reach the brain tumor (glioblastoma) and brain cells (astrocyte) [44].

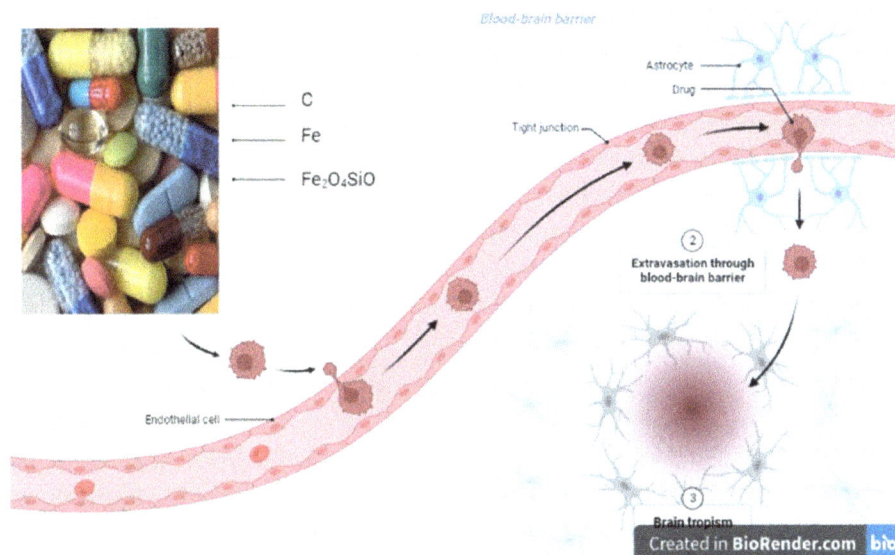

Figure 1.5: Passage of superparamagnetic iron oxide nanoparticles through the BBB and reaching the brain from the blood.

Nanoparticles are widely used as drug delivery agents. Nanoparticles, which can be used as both a lubricant and a drug delivery system, are used instead of surgical procedures in joint diseases such as osteoarthritis [45]. Nanotechnological methods are used in various surgical procedures. There are sensors used to accurately determine blood pressure in the balloon inflation method, which is a technique used to unclog the arteries [46].

Biocidal properties of various nanoparticles are being studied to remove bacteria that threaten public health. NiO nanoparticles, a semiconductor material with magnetic, catalytic and photocatalytic properties, are known to have significant antibacterial properties even at low concentrations. In addition, NiO/chitosan nanocomposites have reported superior antibacterial effects against Gram-negative (*Escherichia coli*) and Gram-positive (*Saccharomyces cerevisiae* and *Bacillus subtilis*) bacteria [47]. ZnO

NPs have antifungal properties and are effective against pathogens such as *Erythricium salmonicolor* [17].

The cosmetics industry uses nanotechnology effectively to realize people's desire to stay beautiful and young. For this, he resolutely pushes his limits. Nano-based materials are applied in various hair care products such as antiaging and skin rejuvenating cosmetics, sunscreen creams and shampoos. For example, nanoemulsions are used in shampoo, conditioner, nail polish, lotion or antiwrinkle creams because they increase skin penetration. Nanoemulsions can deliver ingredients that may be beneficial for the skin into the skin depending on the concentration [48].

1.3.2 Water treatment

Water is nonrenewable and a vital resource for human survival. Due to the increase in population and industrialization, it has led to the pollution of waters [49]. Dyes are important pollutants arising from the wastes of sectors such as food, automotive, textile or cosmetics that harm human health [29, 50]. As an example of these dyes, methyl orange is a carcinogenic azo dye used in the food and textile industry [29]. Methylene blue, a cationic dye, is a dark blue dye with high water solubility. It causes vomiting, diarrhea, tachycardia, shortness of breath and skin irritation as well as harming the environment [9]. Even very small concentrations of dyes contaminating water systems are visible. Due to the coloration of the water, photosynthesis is reduced and therefore the water flora is adversely affected. Plants, animals, the environment and human health are harmed by the pollution of water [9, 40]. In Figure 1.6, nanomaterials used in water treatment and possible contaminants are shown schematically.

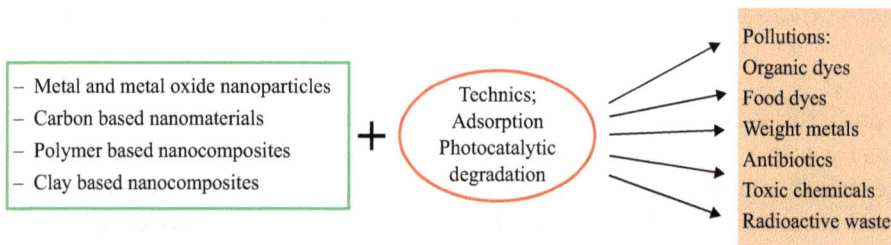

Figure 1.6: Nanoparticles used for cleaning from wastewater and types of pollution in water.

Nanoparticles are widely used in adsorption and photocatalytic degradation applications for water treatment. CuO NPs with large surface area, antibacterial and antifungal properties provide photocatalytic degradation of organic dyes such as methylene blue under UV lamp [9]. Similarly, nanocomposite obtained by using xyloglucan, a biodegradable and nontoxic polysaccharide and magnetic Fe_3O_4, has been reported to be 99% successful in removing cationic dye from aqueous solution when used as a photocatalyst. In order to improve the photocatalyst activity and adsorbent properties of the synthesized nanoparticle, there is a need to continue research with different dyes or pollutants [40].

Mesoporous silica nanoparticles (MSNs) have unique structural properties due to the controllability of their pore diameters. They can also be functionalized with different functional groups. They are used as sorbent in the removal of weight metals such as chromium due to the active sites they have [31]. Materials formed by CNTs containing metal oxide are among the alternatives that can be used to minimize air pollution and solve the water treatment problem [49].

1.3.3 Agriculture

Nanotechnology has the power to change the course of the current food and agriculture industry [39]. The effective use of nanotechnology in agriculture provides benefits in areas such as transport in soil, displacement in plants, and efficient targeting of pests with the use of pesticides at nanoscale, thanks to nanomaterials. New studies such as crop nutrition and smart plant sensors reveal that nanotechnology could be important for sustainable agriculture [51]. Nanobiosensors will help in the fight against undesirable pathogens in the agricultural sector. It is predicted that nanostructured materials that reduce the amount of dose required for crop plants and increase the activity of pesticides will exist in the future [39]. Zinc is beneficial for plant

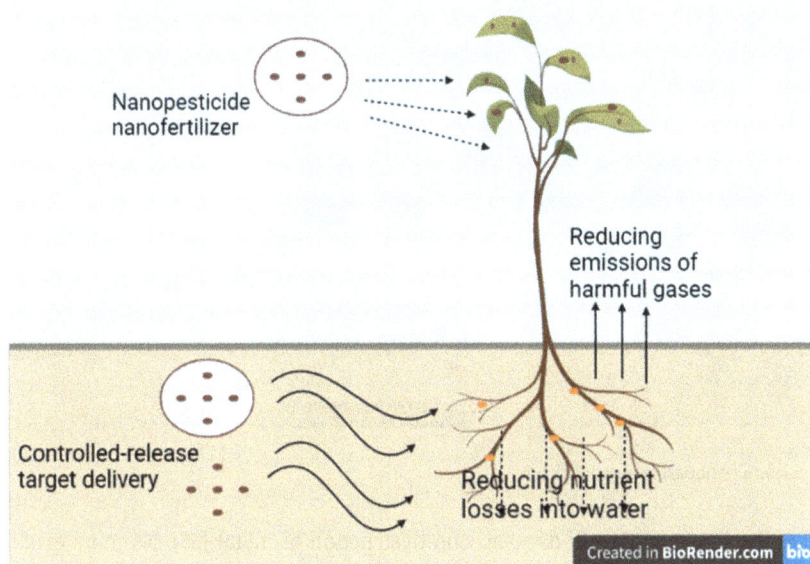

Figure 1.7: The path that nanofertilizers follow in the soil after they affect the plant.

growth and a metabolic regulatory element. In some plants, its deficiency may have consequences such as yellowing and shrinking of the leaves. Processing ZnO nanoparticles into plants by using them in biofertilizers is one of the important studies [52]. Scientific reports report that gold, CeO_2, ZnO, CuO, FeO, Fe/SiO_2, Mg, TiO_2 and calcium nanoparticles are used as nanofertilizers.

The use of nanobiotechnology is very important in terms of minimizing agricultural waste and sustainable development [53]. There are still issues that are not clear. For example, copper, silver and zinc-based nanoparticles have the potential to be used as pesticides and fungicides. However, it is known how this potential will affect the soil quality in the long term. There is a need for the formation of new agricultural models and the development of the obtained data by processing [51]. In Figure 1.7, the effect of the use of nanofertilizers on plants and soil is given visually.

1.3.4 Industrial area

Another use of nanoparticles is as a lubricant. Lubricants have functions such as removing excessive heat, preventing friction, wear and corrosion, and resistance to degradation. It has been reported to provide benefits such as reducing friction and increasing brightness. Properties of nanoparticles such as concentration, size, shape and structure affect tribological properties when added to the lubricant. The movement of CNTs in the lubricant medium is shown in Figure 1.8 [54].

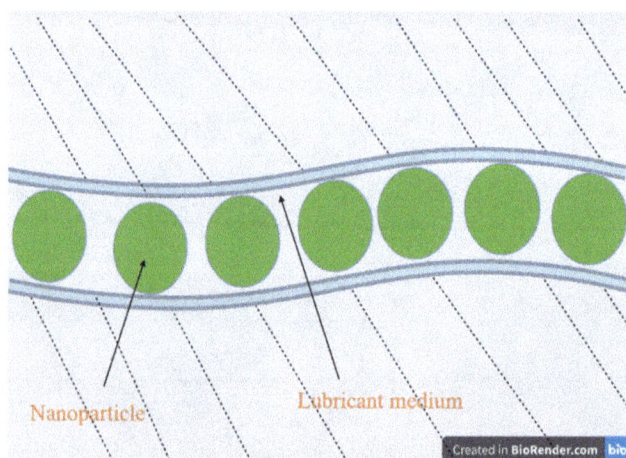

Figure 1.8: Carbon nanoparticles as lubricant.

Corrosion is the irreversible self-damage and destruction of metal [55]. Nanomaterials are unique materials that can show anticorrosive properties due to their surface and volume ratios [1]. The surface/volume ratio of nanoparticles causes them to be effective in corrosion. Despite their small size, they preserve the amount of active site by increasing the surface of a given mass. During inhibition, nanomaterials are subjected to physical or chemical sorption on the metal surface. This result reveals the structure that prevents corrosion [55]. The solubility and sorption abilities of nanomaterials in aqueous electrolytes are important in terms of corrosion [1]. Nanoparticles can increase the corrosion resistance of organic coatings by extending electrolyte paths due to their large surface area. However, if the nanoparticle agglomerates, its efficiency against corrosion may decrease [27]. Silver, silica, iron and copper, ZnO and TiO_2 nanoparticles have anticorrosion ability [55].

Today, many nanoparticles are used to take advantage of the gas sensor properties [56]. Volatile organic compounds (VOC), known to be canserogens, adversely affect human and environmental health. For example, formaldehyde, polycyclic hydrocarbons and some organic waste gases are highly carcinogenic. They can even lead to death from excessive inhalation. Various metal oxide semiconductor nanoparticles and carbon-based materials such as ZnO, SnO_2 and WO_3 show gas sensor properties [57]. Nanocomposite synthesized using ZnO and polyaniline has superior sensor properties for NO_2 gas [56].

Nanotechnology has made progress in electronic applications as well as in all fields. Thanks to their large surface areas, they are candidate materials for flexible electronics. There are many uses such as energy devices, optical and electronic devices, wearable heaters for thermal therapy, sensors, soft electronics and field effect transistors [58, 59]. Copper is an important and inexpensive alternative to noble metals for use in soft electronic circuits. It has high electrical conductivity and is easily available [59].

1.4 The future of nanomaterials

Technological developments in the world now point to the existence of nanotechnology and nanoscience. Scientific data show that when microtechnology is typed into the Google search engine, 130,000 results are obtained, while nanotechnology is written more than 20 million search results [60].

It is expected that the areas affected by developing nanoscience and nanotechnology will increase. Wear-resistant tires, nano-based particles with the best performance of paint pigments, lasers with enhanced performance with new features, nanoparticles to be used in medicine with the ability to deliver the drug to the target, designs of innovative biosensors, production of nano-based traps to eliminate environmental pollution are some of them (Figure 1.9) [61].

As can be seen, the use of nanotechnology and nanomaterials is a subject that develops and comes to the fore in contemporary life. The number of nanoparticles and nanocomposites synthesized by different methods is increasing day by day and their properties are being examined.

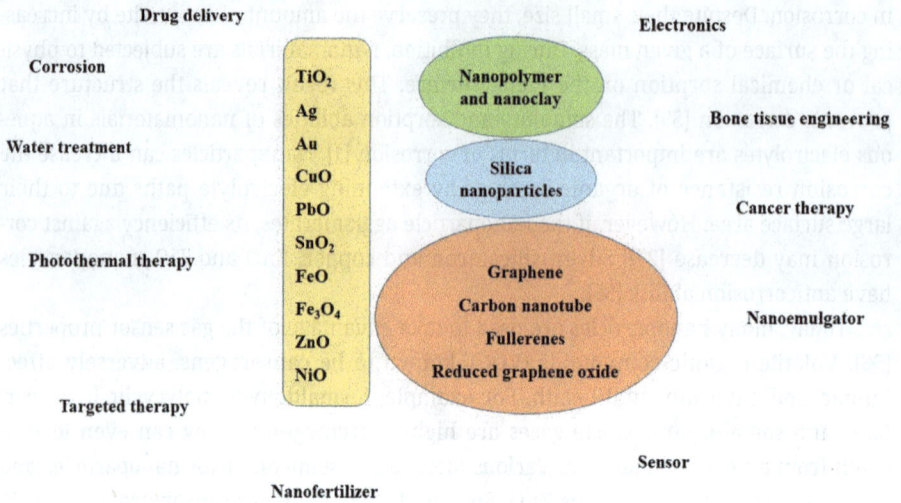

Figure 1.9: Different nanomaterials and applications.

In line with these results,
- it is necessary to understand the physical and chemical structures of many nanomaterials;
- it is important to choose the right application area by investigating its cytotoxic and antibacterial properties as well as its electronic and magnetic properties;
- in addition, the efficient and cost-effective development of synthesis methods is necessary for sustainability.

The fact that nanotechnology can be used in almost all fields is an indication that the scientific studies and industrial applications on this subject will continue rapidly.

Finally, in the next 10 years, nanoscience and nanotechnological applications have been accepted as the most promising scientific and technological development among the technological fields that exist today. Every country in the world has made a large investment in nanotechnological applications. National governments have also led to the emergence of the National Nanotechnology Initiative due to the financing activities in Europe. The major future consequences of nanotechnology on issues such as privacy or the emergence of a common language are alarming. Every society must be prepared for the changes that come with nanoscience and its nanotechnological applications.

References

[1]   Verma, C., Quraishi, M.A., Nanotechnology in the service of corrosion science: Considering graphene and derivatives as examples, Corrosion Engineering Science and Technology, 2022, 57(6), 580–597, doi: 10.1080/1478422X.2022.2093690.

[2]   Dağhan, H., Nano fertilizers, Turkish Journal of Agricultural Research, 2017, 4(2), 197–203, doi: 10.19159/tutad.294991.

[3]   Khan, Z.U.H., Khan, A., Chen, Y., Shah, N.S., Muhammad, N., Khan, A.U., et al., Biomedical applications of green synthesized nobel metal nanoparticles, Journal of Photochemistry and Photobiology B: Biology, 2017, 173(May), 150–164, doi: 10.1016/j.jphotobiol.2017.05.034.

[4]   Ravi Kumar, M.N., Nano and microparticles as controlled drug delivery devices, Journal of Pharmacy & Pharmaceutical Sciences : A Publication of the Canadian Society for Pharmaceutical Sciences, Société Canadienne Des Sciences Pharmaceutiques, 2000, 3(2), 234–258.

[5]   Bhavsar, D., Patel, V., Sawant, K., Systemic investigation of in vitro and in vivo safety, toxicity and degradation of mesoporous silica nanoparticles synthesized using commercial sodium silicate, Microporous and Mesoporous Materials, 2019, 284(January), 343–352, doi: 10.1016/j. micromeso.2019.04.050.

[6]   Jia, Y., Zhang, P., Sun, Y., Kang, Q., Xu, J., Zhang, C., et al., Regeneration of large bone defects using mesoporous silica coated magnetic nanoparticles during distraction osteogenesis, Nanomedicine: Nanotechnology, Biology, and Medicine, 2019, 21, 102040, doi: 10.1016/j.nano.2019.102040.

[7]   Sathya, K., Saravanathamizhan, R., Baskar, G., Ultrasonic assisted green synthesis of Fe and Fe/Zn bimetallic nanoparticles for in vitro cytotoxicity study against hela cancer cell line, Molecular Biology Reports, 2018, 45(5), 1397–1404, doi: 10.1007/s11033-018-4302-9.

[8]   Narayanan, M., Vigneshwari, P., Natarajan, D., Kandasamy, S., Alsehli, M., Elfasakhany, A., et al., Synthesis and characterization of TiO2 NPs by aqueous leaf extract of coleus aromaticus and assess their antibacterial, larvicidal, and anticancer potential, Environmental Research, 2021, 200(January), 111335, doi: 10.1016/j.envres.2021.111335.

[9]   Kütük, N., Çetinkaya, S., Green synthesis of copper oxide nanoparticles using black, green and tarragon tea and investigation of their photocatalytic activity for Methylene Blue Siyah Çay, Yeşil Çay Ve Tarhun Çayı Kullanarak Bakır Oksit Nanoparçacıkların Yeşil Sentezi Ve Metilen, Pamukkale University Journal of Engineering Sciences, 2022, 28(7), 954–962, doi: 10.5505/pajes.2022.47037.

[10]   Chung, H., Yu, M., Nguyen, Q.P., Nanotechnology for oil field applications : Challenges and impact, Journal of Petroleum Science and Engineering Nan, 2017, 157(September 2016), 1160–1169, doi: 10.1016/j.petrol.2017.07.062.

[11]   Tüylek, Z., Story of small things: Nanomaterial, Nevşehir Bilim Ve Teknoloji Dergisi, 2016, 5(2), 130–130, doi: 10.17100/nevbiltek.284737.

[12]   Shahcheraghi, N., Golchin, H., Sadri, Z., Tabari, Y., Borhanifar, F., Makani, S., Nano-biotechnology, an applicable approach for sustainable future, 3 Biotech, 2022, 12(3), 1–24, doi: 10.1007/s13205-021-03108-9.

[13]   Abid, M.A., Kadhim, D.A., Novel comparison of iron oxide nanoparticle preparation by mixing iron chloride with henna leaf extract with and without applied pulsed laser ablation for methylene blue degradation, Journal of Environmental Chemical Engineering, 2020, 8(5), 104138, doi: 10.1016/j. jece.2020.104138.

[14]   Miri, A., Sarani, M., Hashemzadeh, A., Mardani, Z., Darroudi, M., Biosynthesis and cytotoxic activity of lead oxide nanoparticles, Green Chemistry Letters and Reviews, 2018, 11(4), 567–572, doi: 10.1080/17518253.2018.1547926.

[15]   Kütük, N., Mesoporous silica nanoparticles, methods of preparation and use of bone tissue engineering, International Journal of Life Sciences and Biotechnology, 2021, 4(3), 507–522, doi: 10.38001/ijlsb.880711.

[16] Puangjan, A., Chaiyasith, S., Electrochimica acta an Ef Fi cient ZrO 2 / Co 3 O 4 / reduced graphene oxide nanocomposite electrochemical sensor for simultaneous determination of gallic acid, caffeic acid and protocatechuic acid natural antioxidants, Electrochimica Acta, 2016, 211, 273–288, doi: 10.1016/j.electacta.2016.04.185.

[17] Arciniegas-Grijalba, P.A., Patiño-Portela, M.C., Mosquera-Sánchez, L.P., Guerrero-Vargas, J.A., Rodríguez-Páez, J.E., ZnO nanoparticles (Zno-nps) and their antifungal activity against coffee fungus erythricium salmonicolor, Applied Nanoscience (Switzerland), 2017, 7(5), 225–241, doi: 10.1007/s13204-017-0561-3.

[18] Ranjith, K.S., Uyar, T., Polymeric nanofibers decorated with reduced graphene oxide nanoflakes, Biochemical Pharmacology, 2017, October, 4–6, doi: 10.1016/j.mattod.2017.06.006.

[19] Feng, L., Liu, Z., Graphene in biomedicine : Opportunities and challenges special report, Special Report, 2011, 6, 317–324.

[20] Gurunathan, S., Han, J.W., Park, J.H., Eppakayala, V., Kim, J.H., Ginkgo Biloba: A natural reducing agent for the synthesis of cytocompatible grapheme, International Journal of Nanomedicine, 2014, 9 (1), 363–377, doi: 10.2147/IJN.S53538.

[21] Yang, H., Lu, Y., Lin, K., Hsu, S., Huang, C., She, S., et al., Biomaterials EGRF conjugated pegylated nanographene oxide for targeted chemotherapy and photothermal therapy, Biomaterials, 2013, 34 (29), 7204–7214, doi: 10.1016/j.biomaterials.2013.06.007.

[22] Szunerits, S., Boukherroub, R., Szunerits, S., Boukherroub, R., Antibacterial activity of graphene-based materials to cite this version : HAL Id : HAL-01693273, Journal of Materials Chemistry B, 2018, 4(43), 6892–6912, doi: 10.1039/C6TB01647B.

[23] Zuverza-Mena, N., Martínez-Fernández, D., Du, W., Hernandez-Viezcas, J.A., Bonilla-Bird, N., López-Moreno, M.L., et al., Exposure of engineered nanomaterials to plants: Insights into the physiological and biochemical responses-a review, Plant Physiology and Biochemistry, 2017, 110(January), 236–264, doi: 10.1016/j.plaphy.2016.05.037.

[24] Kumar, P., Nano-TiO$_2$ doped Chitosan Scaffold for the bone tissue, International Journal of Biomaterials, 2018, 2018, 1–7.

[25] Pešić, M., Podolski-Renić, A., Stojković, S., Matović, B., Zmejkoski, D., Kojić, V., et al., Anti-cancer effects of cerium oxide nanoparticles and its intracellular redox activity, Chemico-Biological Interactions, 2015, 232, 85–93, doi: 10.1016/j.cbi.2015.03.013.

[26] Nguyen, T.V., Nguyen, T.T.H., Wang, S.L., Vo, T.P.K., Nguyen, A.D., Preparation of Chitosan nanoparticles by TPP ionic gelation combined with spray drying, and the antibacterial activity of chitosan nanoparticles and a chitosan nanoparticle–amoxicillin complex, Research on Chemical Intermediates, 2017, 43(6), 3527–3537, doi: 10.1007/s11164-016-2428-8.

[27] Palimi, M.J., Rostami, M., Mahdavian, M., Ramezanzadeh, B., A study on the corrosion inhibition properties of silane-modified Fe$_2$O$_3$ nanoparticle on mild steel and its effect on the anticorrosion properties of the polyurethane coating, Journal of Coatings Technology and Research, 2015, 12(2), 277–292, doi: 10.1007/s11998-014-9631-6.

[28] Al-Aoh, H.A., Mihaina, I.A.M., Alsharif, M.A., Darwish, A.A.A., Rashad, M., Mustafa, S.K., et al., Removal of methylene blue from synthetic wastewater by the selected metallic oxides nanoparticles adsorbent: Equilibrium, kinetic and thermodynamic studies, Chemical Engineering Communications, 2020, 207(12), 1719–1735, doi: 10.1080/00986445.2019.1680366.

[29] Khwannimit, D., Maungchang, R., Rattanakit, P., Green synthesis of silver nanoparticles using clitoria ternatea flower: An efficient catalyst for removal of methyl orange, International Journal of Environmental Analytical Chemistry, 2020, 00(00), 1–17, doi: 10.1080/03067319.2020.1793974.

[30] Kannan, K., Radhika, D., Vijayalakshmi, S., Sadasivuni, K.K., Ojiaku, A.A., Verma, U., Facile fabrication of CuO nanoparticles via microwave-assisted method: photocatalytic, antimicrobial and anticancer enhancing performance, International Journal of Environmental Analytical Chemistry, 2022, 102(5), 1095–1108, doi: 10.1080/03067319.2020.1733543.

[31] Martin, P., Rafti, M., Marchetti, S., Fellenz, N., MCM-41-based composite with enhanced stability for Cr(VI) removal from aqueous media, Solid State Sciences, 2020, 106(May), 106300, doi: 10.1016/j.solidstatesciences.2020.106300.

[32] Mimouni, R., Askri, B., Larbi, T., Amlouk, M., Meftah, A., Photocatalytic degradation and photo-generated hydrophilicity of methylene blue over ZnO / ZnCr 2 O 4 nanocomposite under stimulated uv light irradiation, Inorganic Chemistry Communications, 2020, 115(December), 107889, doi: 10.1016/j.inoche.2020.107889.

[33] Jayrajsinh, S., Shankar, G., Agrawal, Y.K., Bakre, L., Montmorillonite nanoclay as a multifaceted drug-delivery carrier: A review, Journal of Drug Delivery Science and Technology, 2017, 39, 200–209, doi: 10.1016/j.jddst.2017.03.023.

[34] Guler, C., Gulcemal, S., Guner, A., Akgol, S., Yavasoglu, N.U.K., Polymeric nanoparticles tryptophan-graft-p(HEMA): A study on synthesis, characterization, and toxicity, Polymer Bulletin, 2022, doi: 10.1007/s00289-022-04607-2.

[35] Derman, S., Kızılbey, K., Akdeste, M.Z., Polymeric nanoparticles, Journal of Engineering and Natural Sciences, 2013, 212, 107–120.

[36] Javaid, A., Jalalah, M., Safdar, R., Khaliq, Z., Qadir, M.B., Zulfiqar, S., et al., Ginger loaded polyethylene oxide electrospun nanomembrane : Rheological and antimicrobial attributes, 2022.

[37] Siro, I., Plackett, D., Microfibrillated cellulose and new nanocomposite materials : A review, Cellulose, 2010, 17, 459–494, doi: 10.1007/s10570-010-9405-y.

[38] Haraguchi, K., Nanocomposite hydrogels, Current Opinion in Solid State and Materials Science, 2008, 11(2007), 47–54, doi: 10.1016/j.cossms.2008.05.001.

[39] Rai, M., Ingle, A., Role of nanotechnology in agriculture with special reference to management of insect pests, Applied Microbiology and Biotechnology, 2012, 94(2), 287–293, doi: 10.1007/s00253-012-3969-4.

[40] Ahmad, S., Shankar, S., Mishra, A., Ferromagnetic xyloglucan–Fe3O4 green nanocomposites: Sonochemical synthesis, characterization and application in removal of methylene blue from water, Environmental Sustainability, 2020, 3(1), 15–22, doi: 10.1007/s42398-019-00091-z.

[41] Huang, W.S., Chu, I.M., Injectable polypeptide hydrogel/inorganic nanoparticle composites for bone tissue engineering, PLoS ONE, 2019, 14(1), 1–17, doi: 10.1371/journal.pone.0210285.

[42] El-Deeb, N.M., El-Sherbiny, I.M., El-Aassar, M.R., Hafez, E.E., Novel trend in colon cancer therapy using silver nanoparticles synthesized by honey bee, Nanomedicine & Nanotechnology, 2015, 6(2), 1000265, doi: 10.4172/2157-7439.1000265.

[43] Baykara, O., Kanser Tedavîsînde Güncel Yaklaşimlar, Balikesir Health Sciences Journal, 2016, 5(3), 154–165, doi: 10.5505/bsbd.2016.93823.

[44] Wu, V.M., Huynh, E., Tang, S., Uskoković, V., Brain and bone cancer targeting by a ferrofluid composed of superparamagnetic iron-oxide/silica/carbon nanoparticles (earthicles), Acta Biomaterialia, 2019, 88, 422–447, doi: 10.1016/j.actbio.2019.01.064.

[45] Wei, Q., Fu, T., Lei, L., Liu, H., Zhang, Y., Ma, S., et al., Dopamine-triggered one-step functionalization of hollow silica nanospheres for simultaneous lubrication and drug release, Friction, 2022, 11(3), 410–424, doi: 10.1007/s40544-022-0605-x.

[46] Mamalis, A., Recent advances in nanotechnology, Journal of Materials Processing Technology, 2007, 181, 52–58, doi: 10.1016/j.jmatprotec.2006.03.052.

[47] Vishnuvardhanaraj, G., Biocidal properties of chitosan-encapsulated ternary titanium dioxide-nickel oxide-copper oxide hybrid nanomaterials were prepared via a facile one- pot precipitation process, 2022, 1–9.

[48] Effiong, D.E., Uwah, T.O., Jumbo, E.U., Akpabio, A.E., Nanotechnology in cosmetics : Basics, current trends and safety concerns – a review, 2020, 1–22, doi: 10.4236/anp.2020.91001.

[49] Jain, N., Gupta, E., Kanu, N.J., Plethora of carbon nanotubes applications in various fields – a state-of-the-art-review, Smart Science, 2022, 10(1), 1–24, doi: 10.1080/23080477.2021.1940752.

[50]   Arici, T.A., CTAB/H$_2$O$_2$ modified biosorbent for anionic dye from aqueous solutions: Biosorption parameters and mechanism, Biomass Conversion and Biorefinery, 2021, 0123456789, doi: 10.1007/s13399-021-01920-0.

[51]   Zhang, P., Guo, Z., Ullah, S., Melagraki, G., Afantitis, A., Lynch, I., Nanotechnology and artificial intelligence to enable sustainable and precision agriculture, Nature Plants, 2021, 7(7), 864–876, doi: 10.1038/s41477-021-00946-6.

[52]   Saqib, S., Nazeer, A., Ali, M., Zaman, W., Younas, M., Shahzad, A., et al., Catalytic potential of endophytes facilitates synthesis of biometallic zinc oxide nanoparticles for agricultural application, BioMetals, 2022, 35(5), 967–985, doi: 10.1007/s10534-022-00417-1.

[53]   Shende, S., Rajput, V.D., Gade, A., Minkina, T., Fedorov, Y., Sushkova, S., et al., Metal-based green synthesized nanoparticles: Boon for sustainable agriculture and food security, IEEE Transactions on Nanobioscience, 2022, 21(1), 44–54, doi: 10.1109/TNB.2021.3089773.

[54]   Ali, I., Basheer, A.A., Kucherova, A., Memetov, N., Pasko, T., Ovchinnikov, K., et al., Advances in carbon nanomaterials as lubricants modifiers, Journal of Molecular Liquids, 2019, 279, 251–266, doi: 10.1016/j.molliq.2019.01.113.

[55]   Osheiza, A., Abimbola, A., Popoola, P., Sanni, O., The influence of nanoparticle inhibitors on the corrosion protection of some industrial metals : A review, Journal of Bio- and Tribo-Corrosion, 2022, 8(3), 1–16, doi: 10.1007/s40735-022-00665-1.

[56]   Nguyet, T.T., Duy, L., Thi, Q., Nguyet, M., Thi, C., Dang, X., et al., Novel synthesis of a PANI/ ZnO nanohybrid for enhanced – NO$_2$ gas sensing performance at low temperatures, Journal of Electronic Materials, 2023, 52(1), 304–319, doi: 10.1007/s11664-022-09990-0.

[57]   Li, B., Zhou, Q., Peng, S., Liao, Y., Recent advances of SnO 2 -based sensors for detecting volatile organic compounds, Frontiers in Chemistry, 2020, 8(May), 1–6, doi: 10.3389/fchem.2020.00321.

[58]   Park, J., Hwang, J.C., Kim, G.G., Park, J., Flexible electronics based on one-dimensional and two-dimensional hybrid nanomaterials, InfoMat, 2020, 2(1), 33–56, doi: 10.1002/inf2.12047.

[59]   Feng, Y., Zhu, J., Copper nanomaterials and assemblies for soft, Science China Materials, 2019, 62(11), 1679–1708.

[60]   Munoz-sandoval, E., Trends in nanoscience, nanotechnology, and carbon nanotubes : A bibliometric approach, 2014, doi: 10.1007/s11051-013-2152-x.

[61]   Mansoori, G.A., An introduction to nanoscience & nanotechnology, Nanoscience and Plant-Soil Systems, 2017, 48, 1–16, doi: 10.1007/978-3-319-46835-8.

Omar Dagdag*, Rajesh Haldhar, Seong-Cheol Kim, Walid Daoudi,
Elyor Berdimurodov*, Ekemini D. Akpan, Eno E. Ebenso*

Chapter 2
Polymer nanocomposites: synthesis, modification, properties and applications

Abstract: Nanomaterials (NMs) have received the greatest research attention in the past 10 years because of their unique and multifunctional properties. NMs were employed for both commercial and residential uses. TiO_2 was employed in the field of catalysts, Ag was utilized in the field of sensors, and Al_2O_3 was utilized in the field of antibacterial applications. Nanocomposites (NCs) were a new branch of NMs and they were used in the medicine, chemical industry, automobile and other main industries. The production, functionalization, properties and uses of polymer nanocomposites were discussed and reviewed in this chapter.

Keywords: Nanomaterials, nanocomposites, polymer nanocomposites, one dimension

2.1 Introduction

The polymer nanocomposites (PNCs) are a novel class of advanced engineered materials and they were synthesized to improve the multifunctional properties of polymers using nanoscale fillers. In the present times, PNC materials are employed in numerous areas such as electronics, the automotive industry, architecture, water treatment, packing and biotechnology industries [1–7].

Contribution of the author: Equal contribution by all the authors.

*Corresponding author: Omar Dagdag**, Centre for Materials Science, College of Science, Engineering and Technology, University of South Africa, Johannesburg 1710, South Africa, e-mail: dagdao@unisa.ac.za
*Corresponding author: Elyor Berdimurodov**, Faculty of Chemistry, National University of Uzbekistan, Tashkent 100034, Uzbekistan, e-mail: ebensee@unisa.ac.za
*Corresponding author: Eno E. Ebenso**, Centre for Materials Science, College of Science, Engineering and Technology, University of South Africa, Johannesburg 1710, South Africa,
e-mail: elyor170690@gmail.com
Rajesh Haldhar, Seong-Cheol Kim, School of Chemical Engineering, Yeungnam University, Gyeongsan 38541, Republic of Korea
Walid Daoudi, Laboratory of Molecular Chemistry, Materials and Environment (LCM2E), Department of Chemistry, Multidisciplinary Faculty of Nador, University Mohamed I, Nador 60700, Morocco
Ekemini D. Akpan, Centre for Materials Science, College of Science, Engineering and Technology, University of South Africa, Johannesburg 1710, South Africa

https://doi.org/10.1515/9783111071756-002

Nanocomposite (NC) materials are a combination of two or more components or forms with different structural characteristics. A two-phase material described by the concept "NC" has at least one phase with a 1D in the nanometer (1 nm) range [8]. NC can be categorized into three different classes on the basis of matrix materials: (1) PNC, (2) metal NC and (3) ceramic NC. Melting absorption, sol–gel [9] and polymerization are the various methods for creating PNC [9]. The final characteristics of PNC are significantly influenced by the size and kind of filler material used in polymers. This is due to the diameter of the nanofiller having a significant impact on surface contact with the matrix, adherence, binding, dispersal, etc. As nanoparticle size decreases, these impacts might get worse. For instance, the catalytic activity, adherence, electrical properties and chemical bonding of NPs are drastically altered by a significant surface-to-volume ratio. Additionally, as shown in Figure 2.1, quantification of energies, electromagnetism and molecular movements also started to become active at a smaller nanoscale. Figure 2.2 displays a few illustrations of nanofiller.

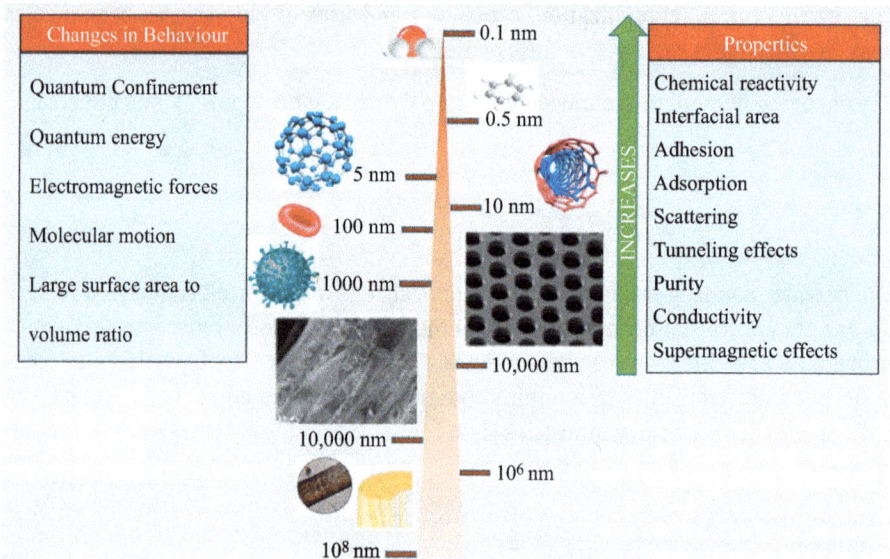

Figure 2.1: Nanoparticle characteristics (reprinted with permission from [10] © 2021 Elsevier Publications).

2.2 Synthesis of PNC

Numerous techniques have been employed to create and produce polymer-based NC. For the purpose of creating high-yield PNC materials, two fundamental categories include direct treatment and ex situ/in situ synthesis.

The most popular form of preparation, direct processing, is used since it is economical and allows for sustained large-scale manufacturing. The top-down method of NM

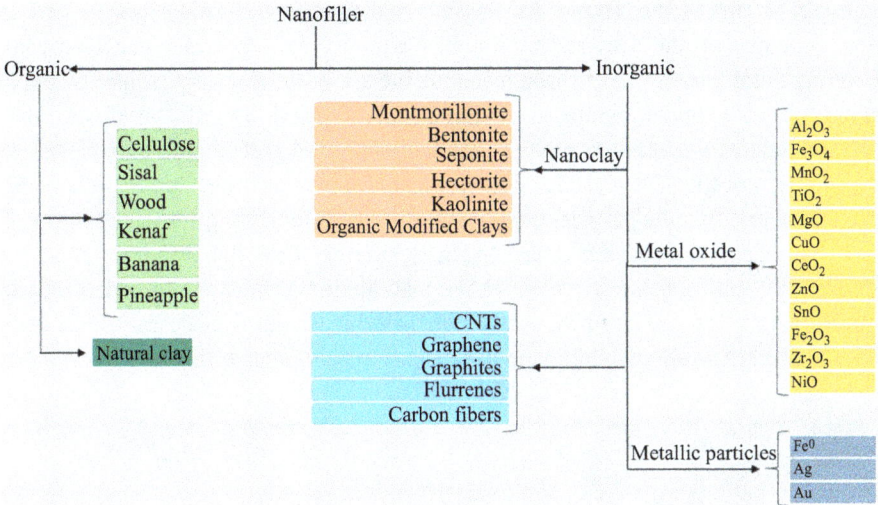

Figure 2.2: Various nanofiller types (reprinted with permission from [10] © 2021 Elsevier Publications).

synthesis includes the direct processing method. Typically, the ensuing hybrids would occur in 1D or 2D, with particles ranging in size from submicron to nanoscale. Here, the polymer and other ingredients are independently synthesized and then combined to form a composite by the use of mechanical mixture, solutions and emulsions. Instead, the method's drawback is the agglomeration and nonhomogeneity of the particles.

However, there are certain tried-and-true methods to get around these issues such as applying dispersants or stabilizers, chemically altering the polymers and other components, and adjusting process settings (time, mixing speed, temperature, etc.). In fact, Pan and Zou [11] created an NC using PAN and tin oxide that had antimony doped into it. Researchers found that the composite formed a conducting channel, and that metallic immobility in the PAN matrix somewhat improved the composite's thermal stability. Researchers recently compared many straightforward compounding techniques. In one recent study, 3D graphene was incorporated into PDMS to serve as a strain sensor [12]. Additionally, methods for creating polymeric interfaces include freeze-drying and ex situ electrospinning.

Ex situ synthesis is carried out using a second method that involves integrating NPs first before distributing them in a polymer solution. The benefit of adopting the ex situ synthesis process is that it allows the user to choose the material without being constrained by the characteristics of the nanofiller and host polymer. As a result, it is not adversely affected by the characteristics of host materials or nanoparticles. However, some significant drawbacks that should be considered during the synthesis process include (1) the potential for nanoparticles to aggregate during blending, (2) the space allocation variable of NP on a polymer matrix, (3) polymer degradation when using melt-compounding technique and (4) disconnection of nanophase from matrix material [10]. Thus, mechanical properties or fusion, solution, or emulsion merge nanofillers and host materials, stabilizing the NP and assisting in avoiding agglomeration [13]. Moreover, to enhance adhesion and dispersion/miscibility among NPs in polymer matrices, compounding conditions can be changed including the use of suitable dispersants or compatibilizers, reactor layout, temperature and shear force [14]. By utilizing a twin-roller mill, melt-mixing was used to produce graphene PNC and graft it onto a polymer [15]. The depiction of the SEM and TEM data in Figure 2.3 demonstrates that there is no NP aggregation and that the particles attain high dispersion and strong adhesive forces [15]. Bao et al. [16] also succeeded in improving the scattering of NPs in polymeric composites by utilizing Figure 2.4's master batch-based melt-mixing technique and their preparation procedure. The produced NC exhibits increased mechanical, electrical conductivity and crystallization rates in addition to having excellent dispersion [16].

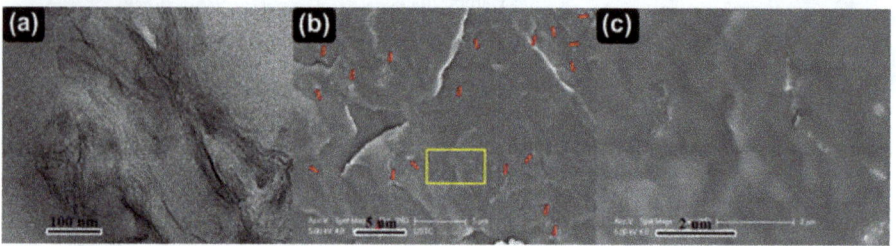

Figure 2.3: TEM photograph of the ultrathin slice in (a), SEM picture of a graphene PNCs surface in (b) and an enlarged SEM image of the chosen rectangular region in (c) (reprinted with permission from [15] © 2014 Elsevier Publications).

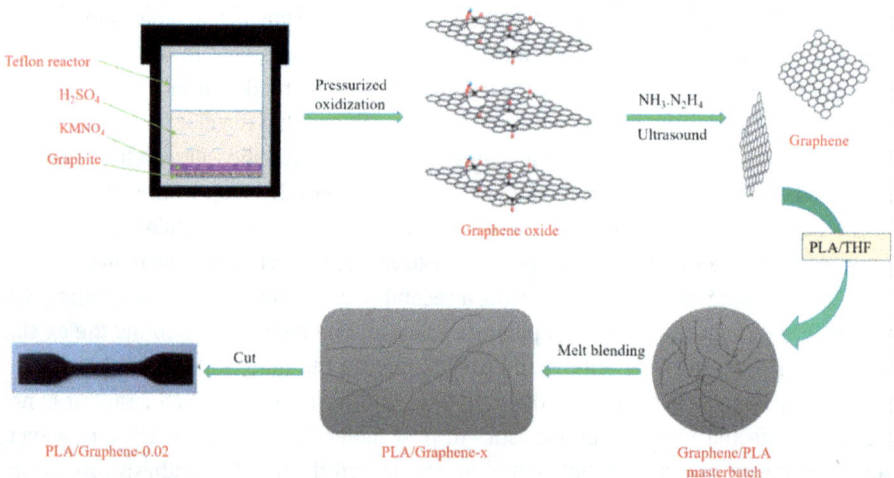

Figure 2.4: Graphene with PLA/graphene NC preparation (reprinted with permission from [10] © 2021 Elsevier Publications).

The third strategy uses a process called "in situ synthesis," which involves creating NPs within a polymer matrix. Polymers stabilize NPs into polymer matrices, which generally operate as nanoreactors and function as a platform for the production of NPs, to solve the problem of NP aggregation [17]. The nature of the NP precursor and functional polymer, the process that produces the NPs, and the content of NPs can all affect the sorts of PNC that can be created utilizing the in situ technique employing functionalized polymer matrix and existing NPs.

Materials like a polymer, precursor, monomer and NP can be used to create NC utilizing in situ synthesis (they are often the initial component) regarding the method of manufacturing. Figure 2.5 illustrates the classification of various in situ NC preparation techniques into three categories [18].

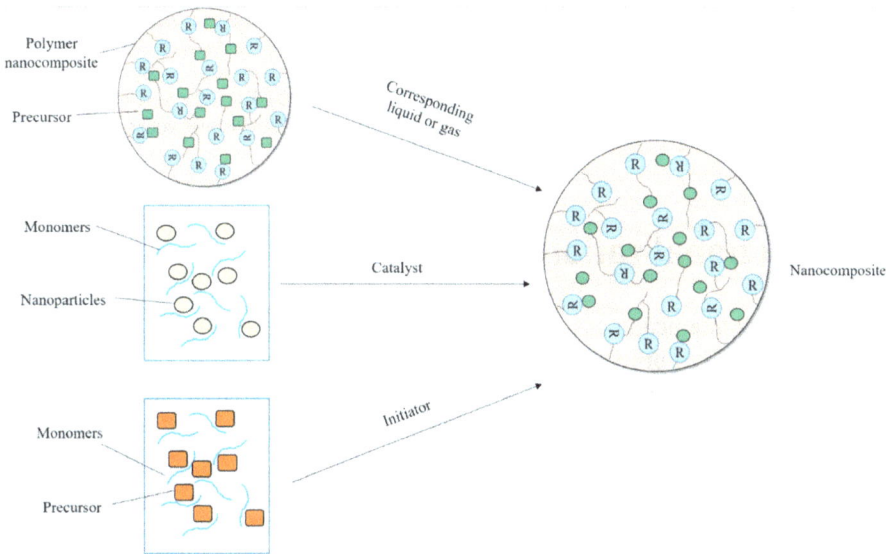

Figure 2.5: The in situ synthesis process of PNCs (reprinted with permission from [18] © 2011 Elsevier Publications).

The fourth strategy involves mixing the NP and polymer precursors with the right solvent [19, 20]. For instance, the sol–gel method is used to create PIS-TiO$_2$ hybrid NC films, which increases chemical bonds at the inorganic and organic interface [21]. It was discovered that TiO$_2$-NPs, with a mean size compared to fewer than 50 nm, were disseminated extremely efficiently in PIS matrix. Figure 2.6 shows how the sol–gel method was used to create MWNTs-TiO$_2$ NC. SEM pictures of the surface of the MWCNT/TiO$_2$ NC revealed that the TiO$_2$ particles were uniformly deposited but that there was considerable agglomeration along the MWCNT [22]. This might be caused by exposed CNT surfaces.

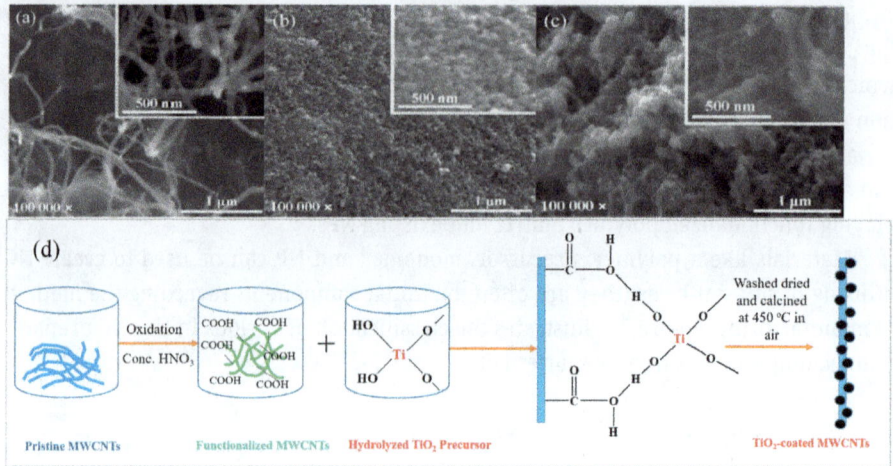

Figure 2.6: (a, b, c) MWCNT, TiO₂ and MWCNT/TiO₂ NC surfaces as seen by SEM; (d) a schematic showing the creation of MWCNT-TiO₂ NC (reprinted with permission from [22] © 2014 Elsevier Publications).

2.3 Functionalization of PNC

To increase the interaction between phases of the nanofiller and the polymer matrix, functionalization is a procedure that modifies the surface of the nanofiller. To maximize the transfer of load across their interfaces, it is also essential to establish uniform dispersion of the nanofiller and generate a strong interfacial connection here between the polymer and the nanofiller [23]. Due to strong Van der Waals forces, dipole–dipole contact and other factors, the unaltered nanoparticle may occasionally agglomerate, reducing its aspect ratio and eroding its traditional features. Therefore, it is crucial to alter the exterior of NPs to preserve their exceptional capabilities. Covalent complexation and noncovalent are the two categories into which these functionalization techniques are separated.

The process of covalent functionalization involves altering the bond's state, graphene and CNT, for instance, functionalization converts their sp^2 carbon atoms to sp^3 carbon atoms. Either "end and flaw functionalization" or "side-wall functionalization" is used to carry out this functionalization. By using esterification and oxidation procedures to modify the R-CO₂H group's surface on the defect sites of CNTs, end defects were functionalized [24] as opposed to the sidewall, which entails the inclusion of chemical agents such as cyclones or radicals and the direct attachment of functional groups onto to the CNT structure. H_2SO_4, HNO_3 and $KMnO_4$ have often utilized oxidants for the terminal and defect functionalization that produced the –CO₂H group, whereas O_3 and H_2O_2 are moderate oxidants [25]. These oxidizing substances provide a uniform dispersion of the nanofiller by reducing the Van der Waals contact between the polymer and nanofiller [26]. The nitric acids oxidizing functionalization approach were used by Jun et al. [26] to eliminate metal particle contaminants from virgin CNTs. In a different study, it was shown that acid functionalization prevented sepio-lite clay (nanofiller) from aggregating in the polymer matrix and reduced matrix-particle interaction compared to nonfunctionalized NC [27]. Additionally, condensation reaction, electrophilic addition, nucleophilic substitution, addition, and are other methods for covalently functionalized graphene.

Well after the use of surfactant [28], polymer wrapping [29] or organic solvents, the noncovalent functionalization technique enhances solubility and biocompatibility without altering CNT's π–π conjugation. Additionally, upon functionalization, the graphene structure or CNTs as well as their physical characteristics (optical and electrical characteristics) remain unchanged [30]. For noncovalent functionalization, common polymers and surfactants include polyvinyl, polysulfone, polyaniline [31], sodium dodecyl sulfate (SDS) and gum Arabic. Due to the lesser adhesion contact between the nanofillers, noncovalent functionalization processes are unstable and weaker than covalent functionalization.

2.4 Characterization of PNC

To identify their mechanical, molecular and chemical properties, the polymeric interfaces created by various methods are analyzed. Additionally, only a small number of novel analytical techniques are used after the original use. Important analytical techniques including XPS, NMR, FTIR, XRD and MS are covered in this section. First, microscopic methods such as FESEM and TEM are used to examine the physical form and orientation of the generated polymer interface. Second, using XRD, the structural orientation of the various components is detected and compared to the normative patterns. Similarly, NMR is applied to the composite to confirm its structural properties. Here, the carbon-based polymeric NMs are revealed by the ^{13}C-based resonance, while the molecular chemistry of the NC is captured by both ^{13}C and ^{1}H solid-state spectroscopic investigation. The chemical makeup as well as oxidation states of the components that make up the polymer interface is then determined by XPS. Significantly, the XPS allows for visualization of the composite's electrical configuration. The FTIR is also used to identify the various functional groups that are present at the polymer interaction or to identify active adsorption sites.

2.5 Properties of PNC

2.5.1 Electrical properties

The addition of CNTs to composites considerably improves the composite's electrical conductivity. Additionally, it is claimed that the use of CNTs enhances the mechanical and thermal characteristics of composites. Electrical and thermal characteristics are considerably improved by multiscale reinforcement with CNTs. Standard fillers, such as carbon and glass fibers, offer a potential approach to the creation of composites with many uses [32]. Numerous fiber-reinforced composites and clean thermoplastics can be considerably improved by adding various NPs such as CNTs. The conductance of the polymeric layer and organic sheet has improved as a result of the inclusion of CNTs and carbon black. Several variables, including the proportion of conductive fillers employed in the polymers, affect electrical conductivity. Compared to amorphous polycarbonate, semicrystalline polyamide has lower electrical conductivity.

The aggregation of nanofibrils even during the conversion of polymer films into organic sheets increases electrical conductivity [33].

2.5.2 Mechanical properties

Prior until now, nanocellulose matrices were employed with thermoplastic matrices, demonstrating the advantages of high toughness and recyclability. The researchers in this study spoke about certain mechanical characteristic findings for nanocellulose thermoset composites. Using nanocellulose in thermoplastic composites greatly improves the strength and stiffness of the material. Nanocellulose-based dispersed or particulate composites benefit from the resin's contact with the comparatively high aspect ratio of the cellulose fibers or particles. Additionally, it has been observed that the influence of fiber contents on the mechanical and thermal expansion characteristics of biocomposites developed on CNF. Using phenolic resin, it has been found that fiber content can improve laminarly by up to 40%. CNF was added to the epoxy, and the glassy retention modulus was increased. At 30 °C and 5% CNF epoxy layer, the modulus increases from 2.6 GPa toward 3.1 GPa. A significant rise is noted above the T_g the stretchy storage modulus for 5% weight CNF's epoxy coating at 130 °C rises from 9.7 MPa through 37.3 MPa [34]. According to Ruiz et al., introducing CNF at approximately 2% weight significantly improves the mechanical characteristics of composite materials, while adding more CNF diminishes those capabilities owing to agglomeration. Okubo et al. investigated how CNF reinforcement was distributed around bamboo strands in the PLA matrix, and they discovered that the inclusion of CNF enhances the fracture characteristics, preventing unexpected crack formation. When CNF is introduced, there is a 200% rise in fracture energy. Researchers have looked at how properly processing CNF might result in lightweight composites with unique barriers and transparent properties that have been used in energy storage systems, drugs, the production of automobiles, sensors, packing and electronics. Films made of nanocellulose can be utilized as barriers. Due to their hydrophobic properties, high-porosity aerogels may also be utilized to collect moisture while allowing the movement of gases [35].

2.5.3 Optical properties

The impact of particles treatments on the optical characteristics of the NC was examined by Kleissl et al. A superior optical transparency value in the near-infrared is produced by in situ sanitizing nanosized aluminum oxide (Al_2O_3) mixed in MMA matrix and by subsequent polymerization as opposed to untreated aluminum oxide [36]. Abdelrazek et al. [37] investigated adding gold nanoparticles (Au NPs) to PEO/PVP composites. The findings showed that adding more Au NPs to the composite significantly increased the values of optical parameters such as Urbach energy, refractive index and optical energy gap.

2.5.4 Magnetic properties

One form of composite with metal NPs and another with ferrite NPs exhibit magnetic characteristics. The majority of the nanoparticles lack hysteresis, which suggests superparamagnetic substance. The PNC includes 2.8% of Fe_2O_3 concentration, according to Ziolo et al. They discovered that these matrices are optically transparent and devoid of hysteresis at room temperature. They also discovered that NC is devoid of hysteresis in electromagnetic polymer materials that include γ-Fe_2O_3 NPs [38]. Nickel oxide nanoparticles (NPs) were integrated into polyvinyl cinnamate, and their magnetic characteristics were studied by Ramesan et al. They discovered that PNC is ferromagnetic. Furthermore, the addition of nickel oxide NPs to polymer was shown to boost remanence, magnetic responsiveness and hysteresis values [39].

2.6 Application of PNC

2.6.1 Food industries

For numerous applications, including food packaging that demands good mechanical, antibacterial and barrier qualities, Ag NPs were added to an HPMC matrix [40]. The samples with smaller Ag NPs produced the greatest results. Additionally, produced films showed excellent antibacterial properties versus *S. aureus* and *E. coli*. Ag NPs decreased the water vapor permeability of NC films and enhanced their mechanical properties (from 28.3 MPa for pure HPMC to 51 MPa for HPMC/Ag). These characteristics showed suggested that HPMC/Ag NC films may be utilized effectively for food packaging.

2.6.2 Solar cells

The amount of energy consumed today is quite significant, and antiresources such as fossil fuels deplete as a result, hence the importance of renewable energy. Nanotechnology is a rapidly developing technology with numerous uses across all industries. Due to their unique qualities, PNC is a novel difficult material with a wide range of applications. Polymers enclosed in nanometals can raise the solar cell's efficiency. It can more precisely capture solar radiation. The best alternative for solar cells is plasmonic metal-NPs since their quantum yield value is larger than one. MPCs, or metal-polymer composites, function like metal-dielectric combinations. SPR stands for the aggregate oscillations of electron clouds. This SPR was produced at the intersection of the metal NPs and the dielectric matrix. This oscillation is caused by the interactions between the metal NPs and the incident light [41].

2.6.3 Sensors

Sensors are employed to identify environmental changes. They, therefore, have biological uses in addition to environmental ones. Because of the SPR excitation, PNC serves as a sensor [42]. The SPR is one of metal NPs' most crucial properties. Applications across several disciplines are provided by the combination of metal and polymer characteristics. The MPC can boost sensors' effectiveness. The SPR oscillations provide the sensor with its sensing ability. Therefore, these matrices are utilized in biosensors as virus and cancer molecule detectors [43]. In sensor-oriented applications, PNC is used to analyze biological and environmental problems.

2.6.4 Thin films

Different methods are being employed to create thin films. Metal NP thin-film fabrication from a single material is challenging. The creation of enclosed thin films and a substrate for metal-NC is assisted by polymers. The creation or modeling of nanodevices requires the use of polymers containing metal NPs. When combined with polymers, metal nanoparticles can make higher-quality films than when used alone. These films are divided at the microphase level between the polymer matrix and the metal

NPs. Strong covalent connections and light weights molecular ligands join the polymer and NPs. Thin films are utilized to produce the SPR-based synthesis sensors [43].

2.6.5 Microbial

Silver nanoparticle and PVA thin-film combination exhibit antibacterial action. They may be reused and are simple to use for film monitoring [44]. PVA and other polymers have antimicrobial properties. We can do better. By combining inorganic NPs with elements like Cu and Ag, antibacterial properties can be enhanced. The discharge of metal ions into the microbial cell aids in the research of their antibacterial properties. Metal ions that have been discharged have the potential to harm cells' microenvironments [45, 46]. Excellent photobactericidal activity is provided by the TiO_2 polymer matrices containing inorganic nanofillers.

2.6.6 Conductance

The conducting quality can be improved using polymer metal NC. To include it in electrical circuitry, the electro-spinning technique may be used to create nanofibers. This can increase conductivity when polymers with conducting properties are created. These polymers are utilized due to their flexibility, affordability, lightweight and excellent solvent reactivity. In the realm of biomedical engineering, they are electroactive biomaterials [47, 48]. Depending on the number of free electrons, a substance's conductivity will vary. Due to the unbound electrons in the atoms, a polymer is electrically conductive. Conjugated double bonds are a kind of polymer structure where single and double bonds are present alternatively. Due to oxidation or reduction, the structure of a polymer with conjugated double bonds can become conductive. Different particle doping can increase conductivity [49]. Photodiodes in electroluminescence usage, LEDs and energy-saving bulbs are all made of semiconductor polymers [50].

2.6.7 Treatment of wastewater

Several hybrids, cutting-edge materials, including nanoclay/conducting polymers, have been developed from polymer/clay NCs that have qualities superior to those of traditional composite materials. Depending on the modification process, a new material composed of polymer (blended) as well as inorganic clay minerals can produce NC. The resultant NC has clay minerals intercalated in the interlayer gaps of the clay, adsorbed on the surface and disseminated throughout the polymer matrix [51]. Intercalated and exfoliated composites of polymer and nanoclay predominate, as seen in Figure 2.7a and b.

The separation of a single layer of clay results in an unorganized clay dispersal in the polymer matrix, whereas intercalated NCs are created by the orderly complexation of the polymer matrices in the interlayer region of layered clay [52]. The elimination of inorganic contaminants from an aqueous environment has been the most extensively researched polymer/nanoclay composite. Tetramethylethylenediamine (TMEDA), functioning as the initiator and accelerator, and potassium persulfate ($K_2S_2O_8$) were present when the clay/PMEA was synthesized utilizing the bulk polymerization approach [53]. Lead(II) ions are removed from the aqueous environment using PMEA, which has an adsorption capability of around 81.02 mg/g. According to the batch adsorption investigation, adsorption is not affected by the pH range of 4–6, temperature, time and C_0 all have a role, though. Lead ion was adsorbing onto clay/PMEA NC in an endothermic process that involved physisorption. To regenerate clay/PMEA NC 10 mM nitric acid, which may be utilized again after 5 cycles of both adsorption and desorption, was employed. Additionally, lead ions were removed using PAA/BT NC, which has an adsorption capability of 93 mg/g [54]. As illustrated in Figure 2.7c, exfoliated PP is organically changed to create montmorillonite clay NC by in situ crosslinking of pyrrole monomer for the elimination of Cr (VI) from aqueous medium.

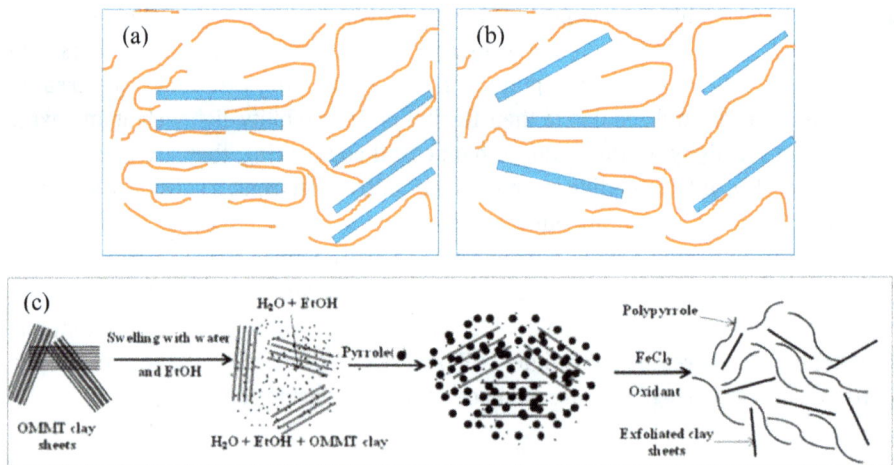

Figure 2.7: Illustration of polymer/clay nano – composites (a) agglomerated, (b) exfoliated and (c) production of highly porous PP-organically transformed montmorillonite clay NC (reprinted with permission from [55] © 2013 Elsevier Publications).

2.6.8 Biomedical

Numerous polymers, including proteins and synthetic ones like PHB, PLA, PCL and PGA, as well as natural polymers like chitin, alginate, chitosan and polysaccharides, can be used in biomedical applications [56]. They have a controlled rate of breakdown, are biocompatible and degrade into nontoxic components. Additionally, the water-soluble polymers PVA, PAA, PEG and guar-gum were utilized in this sector [56]. Additionally, conductive polymers including PP, PVP and PANI were used. Additionally, several researches reported the usage of additional polymers for biomedical purposes including polypropylene, PU and PE [56].

2.6.9 Membranes

Membranes made of polymers or modified polymers are crucial for environmental cleanup. The creation of polymer nanointerfaces-based membranes enhances their permeability, selectivity, mechanical strength and other capabilities by specific physical or chemical alterations in their matrix. Concurrently, this would make up for the conventional shortcomings such as biofouling, low flow and inadequate mechanical resistance [101]. In order to increase efficiency, PNC are often used as nanosized fillers in polymer matrices (membrane substrate) (Figure 2.8).

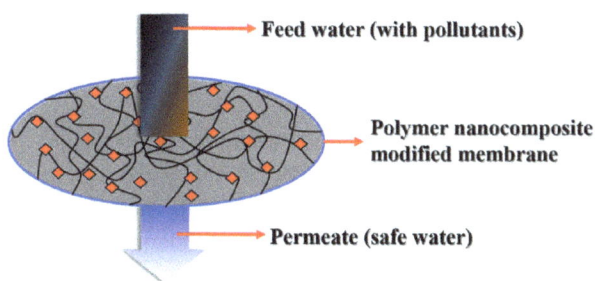

Feed water (with pollutants)

Polymer nanocomposite modified membrane

Permeate (safe water)

Figure 2.8: Application of membranes in water treatment (reprinted with permission from [57] © 2021 Elsevier Publications).

2.6.10 Photocatalysis

PNCs have been widely employed as catalysts in heterogeneous catalysis [58]. PNCs, however, are being used for their effective photo- and electrocatalytic reactions to eliminate difficult-to-remove contaminants from water and waste. As of now, a wide range of synthetic polymers such as PP, PANI, polypropylene and PDMS are combined with semiconductor materials to form photocatalysts. The example of the TiO_2-integrated polymeric composite is valuable for understanding the behavior of polymeric interface in photocatalytic applications. Due to its usefulness in collecting UV and visible light, chemical stability, quick conveyance of photo-induced charges, and conjugated polymer is utilized in this situation. TiO_2 is also known to have a high photoelectric conversion efficiency. Therefore, within 90 min, the PANI/wide band-gap semiconductors (TiO_2)

demonstrated about 80% the methylene blue dye photodegradation [58]. Figure 2.9 shows a typical photocatalytic deterioration.

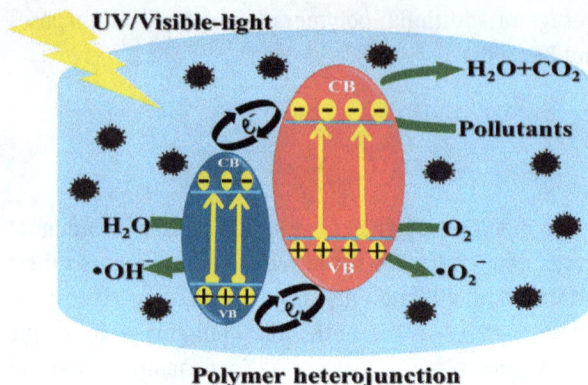

Figure 2.9: Pollutant degradation via typical photocatalysis employing polymer interfaces (reprinted with permission from [57] © 2021 Elsevier Publications).

2.6.11 Energy storage

Polymer NCs have greatly aided in the development of renewable energy as well as its storage, which is one of humanity's most pressing demands. PNCs have demonstrated their promise in several electrochemical energy storage systems [59]. The supercapacitors' electrochemical reaction is carried out with exceptional efficiency, thanks to the increased active surface area provided by the conducting polymer surfaces. Similar to this, PNC and pristine polymers with porous structures in batteries inhibit the mobility of extraneous ions other than Li^+ through all the separators, increasing the batteries' columbic efficiency [60].

Supercapacitors have a greater capacitance than regular capacitors, and they store electrochemical energy through redox processes on the electrode's surface or at the electrode-electrolyte interface [61]. Due to its distinct qualities including quick charge/discharge, cycle endurance, power density and high energy, supercapacitors are the forthcoming energy storage devices. Polythiophene, PANI and PPy are three conducting polymers that are currently used in the production of supercapacitors. To construct the electrodes, PNC are often combined with metal oxides or carbon allotropes, improving the electrochemical activity. In particular, the development of polymer interfaces lessens the deformation of polymeric chains while enhancing the cyclic durability and rate capability of supercapacitors [62].

The most potential and popular high-energy-density electrochemical energy storage technology is represented by Li-ion batteries. Performance and formability have improved with Li-ion battery technology throughout time, thanks to a variety of NMs.

Rechargeable batteries frequently employ PNC-based Li-ion batteries due to its benefits such as lower toxicity, more capacity, higher operating voltage and better cycle
life [63]. Traditionally, the anode materials used in these batteries include silica,
graphite and other metal oxides, whereas the cathode materials are ternary composites such as $LiFePO_4$, $LiMn_2O_4$ and $LiCoO_2$.

Applications for metal oxide NC are shown in Figure 2.10.

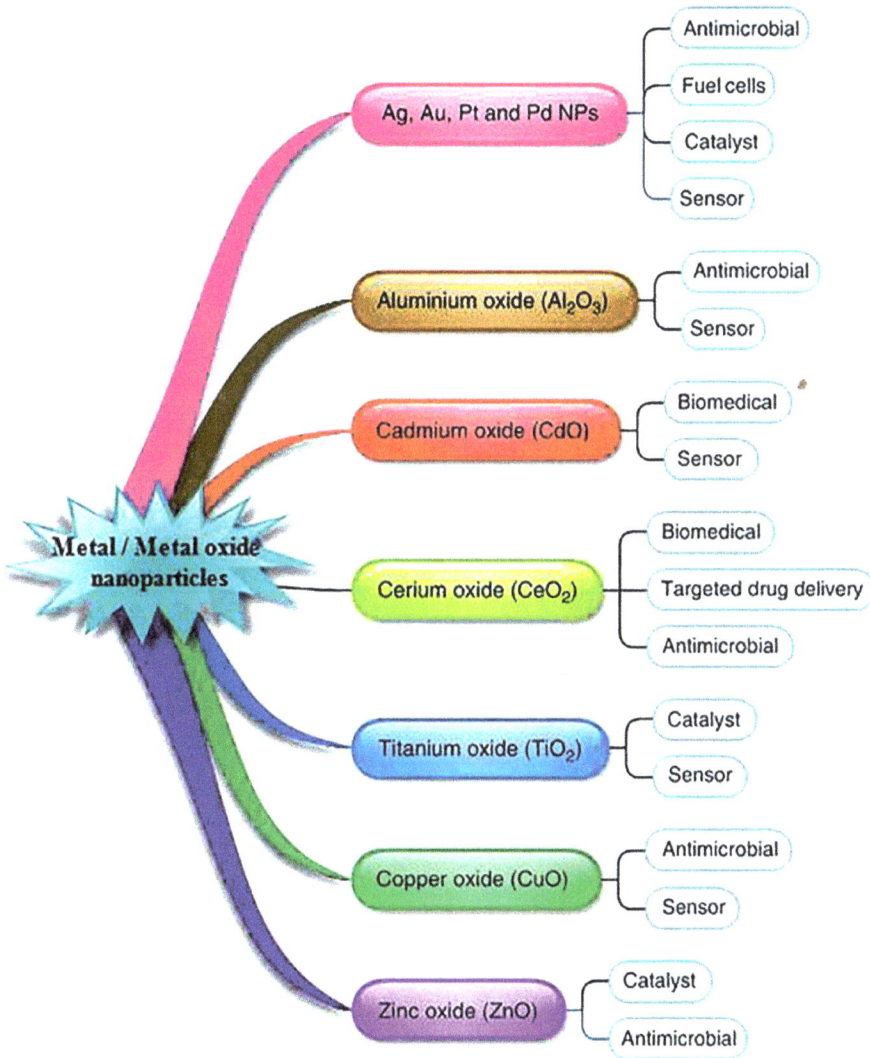

Figure 2.10: Applications of metal oxide NC (reprinted with permission from [64] © 2021 Elsevier
Publications).

2.7 Conclusions

Nanotechnology is a rapidly expanding field of study because of its numerous applications in physical, chemical, biological science, etc. It is confirmed that the NC-based materials are 1D related to their size, structural properties, mechanical and other performances. The NC materials were categorized into three types related to their matrix structure. PNC is the main type of NC material, which is designed with a polymer matrix that contains nanoscopic components. Researchers have reported using a variety of techniques to create this polymer-encapsulated NC. There are several applications for them because of their low weight, downward pressure, flexibility and other characteristics.

Abbreviations

NMs	Nanomaterials
PNC	Polymer nanocomposites
NC	Nanocomposites
1D	One dimension
NPs	Nanoparticles
PIS	Poly(imide siloxane)
MWNTs	Multiwalled carbon nanotubes
CNT	Carbon nanotube
PLA	Polylactic acid
MMA	Methyl methacrylate
NIR	Near-infrared reflectance
PEO/PVP	Polyethylene oxide/polyvinyl pyrrolidone
HPMC	Hydroxyl-propyl methylcellulose
SPR	Surface plasma resonance
PVA	Poly(vinyl alcohol)
LEDs	Light-emitting diodes
PMEA	Poly(methoxyethyl acrylamide)
PAA/BT	Poly(acrylic acid)/bentonite
PP	Polypyrrole
PLA	Poly(lactic acid)
PCL	Poly(ε-caprolactone)
PGA	Poly(glycolic acid)
PHB	Poly(hydroxyl butyrate)
PVA	Poly(vinyl alcohol)
PAA	Poly(acrylic acid)
PEG	Poly(ethylene glycol)
PNA	Polyaniline (PANI)
PVP	Polyvinylpyrrolidone
PE	Polyethylene
PU	Polyurethane
PAN	Polyacrylonitrile
PDMS	Polydimethylsiloxane

References

[1]   Belgaonkar, M.S., Kandasubramanian, B., Hyperbranched polymer-based nanocomposites: Synthesis, progress, and applications, European Polymer Journal, 2021, 147, 110301.

[2]   Zinge, C., Kandasubramanian, B., Nanocellulose based biodegradable polymers, European Polymer Journal, 2020, 133, 109758.

[3]   Gore, P.M., Balakrishnan, S., Kandasubramanian, B., Superhydrophobic corrosion inhibition polymer coatings, in Superhydrophobic Polymer Coatings, Elsevier, 2019, 223–243.

[4]   Malik, A., Kandasubramanian, B., Flexible polymeric substrates for electronic applications, Polymer Reviews, 2018, 58, 630–667.

[5]   Badhe, Y., Balasubramanian, K., Novel hybrid ablative composites of resorcinol formaldehyde as thermal protection systems for re-entry vehicles, RSC Advances, 2014, 4, 28956–28963.

[6]   Yadav, R., Goud, R., Dutta, A., Wang, X., Naebe, M., Kandasubramanian, B., Biomimicking of hierarchal molluscan shell structure via layer by layer 3D printing, Industrial & Engineering Chemistry Research, 2018, 57, 10832–10840.

[7]   Gore, P.M., Naebe, M., Wang, X., Kandasubramanian, B., Progress in silk materials for integrated water treatments: Fabrication, modification and applications, Chemical Engineering Journal, 2019, 374, 437–470.

[8]   Okpala, C.C., Nanocomposites–an overview, International Journal of Engineering Research and Development, 2013, 8, 17–23.

[9]   He, J.-P., Li, H.-M., Wang, X.-Y., Gao, Y., In situ preparation of poly (ethylene terephthalate)–SiO2 nanocomposites, European Polymer Journal, 2006, 42, 1128–1134.

[10]   Ahmad, K., Study of different polymer nanocomposites and their pollutant removal efficiency, Polymer, 2021, 217, 123453.

[11]   Pan, W., Zou, H., Characterization of PAN/ATO nanocomposites prepared by solution blending, Bulletin of Materials Science, 2008, 31, 807–811.

[12]   Wu, S., Ladani, R.B., Zhang, J., Ghorbani, K., Zhang, X., Mouritz, A.P., et al., Strain sensors with adjustable sensitivity by tailoring the microstructure of graphene aerogel/PDMS nanocomposites, ACS Applied Materials & Interfaces, 2016, 8, 24853–24861.

[13]   Chan, C.-M., Wu, J., Li, J.-X., Cheung, Y.-K., Polypropylene/calcium carbonate nanocomposites, Polymer, 2002, 43, 2981–2992.

[14]   Yu, Q., Wu, P., Xu, P., Li, L., Liu, T., Zhao, L., Synthesis of cellulose/titanium dioxide hybrids in supercritical carbon dioxide, Green Chemistry, 2008, 10, 1061–1067.

[15]   Yuan, B., Bao, C., Song, L., Hong, N., Liew, K.M., Hu, Y., Preparation of functionalized graphene oxide/polypropylene nanocomposite with significantly improved thermal stability and studies on the crystallization behavior and mechanical properties, Chemical Engineering Journal, 2014, 237, 411–420.

[16]   Bao, C., Song, L., Xing, W., Yuan, B., Wilkie, C.A., Huang, J., et al., Preparation of graphene by pressurized oxidation and multiplex reduction and its polymer nanocomposites by masterbatch-based melt blending, Journal of Materials Chemistry, 2012, 22, 6088–6096.

[17]   Sahoo, B.P., Tripathy, D.K., Introduction to clay-and carbon-based polymer nanocomposites: Materials, processing, and characterization, in Properties and Applications of Polymer Nanocomposites, Springer, 2017, 1–24.

[18]   Zhao, X., Lv, L., Pan, B., Zhang, W., Zhang, S., Zhang, Q., Polymer-supported nanocomposites for environmental application: A review, Chemical Engineering Journal, 2011, 170, 381–394.

[19]   Li, X., Wang, D., Cheng, G., Luo, Q., An, J., Wang, Y., Preparation of polyaniline-modified TiO_2 nanoparticles and their photocatalytic activity under visible light illumination, Applied Catalysis B: Environmental, 2008, 81, 267–273.

[20] Utracki, L., Sepehr, M., Boccaleri, E., Synthetic, layered nanoparticles for polymeric nanocomposites (PNCs), Polymers for Advanced Technologies, 2007, 18, 1–37.

[21] Liaw, W.-C., Chen, K.-P., Preparation and characterization of poly (imide siloxane) (PIS)/titania (TiO2) hybrid nanocomposites by sol–gel processes, European Polymer Journal, 2007, 43, 2265–2278.

[22] Abd Hamid, S.B., Tan, T.L., Lai, C.W., Samsudin, E.M., Multiwalled carbon nanotube/TiO_2 nanocomposite as a highly active photocatalyst for photodegradation of Reactive Black 5 dye, Chinese Journal of Catalysis, 2014, 35.

[23] Punetha, V.D., Rana, S., Yoo, H.J., Chaurasia, A., McLeskey Jr, J.T., Ramasamy, M.S., et al., Functionalization of carbon nanomaterials for advanced polymer nanocomposites: A comparison study between CNT and graphene, Progress in Polymer Science, 2017, 67, 1–47.

[24] Lau, Y.J., Khan, F.S.A., Mubarak, N., Lau, S.Y., Chua, H.B., Khalid, M., et al., Functionalized carbon nanomaterials for wastewater treatment, Industrial Applications of Nanomaterials, 2019, 283–311.

[25] Liang, S., Li, G., Tian, R., Multi-walled carbon nanotubes functionalized with a ultrahigh fraction of carboxyl and hydroxyl groups by ultrasound-assisted oxidation, Journal of Materials Science, 2016, 51, 3513–3524.

[26] Jun, L.Y., Mubarak, N., Yon, L.S., Bing, C.H., Khalid, M., Abdullah, E., Comparative study of acid functionalization of carbon nanotube via ultrasonic and reflux mechanism, Journal of Environmental Chemical Engineering, 2018, 6, 5889–5896.

[27] Bilotti, E., Fischer, H., Peijs, T., Polymer nanocomposites based on needle-like sepiolite clays: Effect of functionalized polymers on the dispersion of nanofiller, crystallinity, and mechanical properties, Journal of Applied Polymer Science, 2008, 107, 1116–1123.

[28] Erol, O., Uyan, I., Hatip, M., Yilmaz, C., Tekinay, A.B., Guler, M.O., Recent advances in bioactive 1D and 2D carbon nanomaterials for biomedical applications, Nanomedicine: Nanotechnology, Biology and Medicine, 2018, 14, 2433–2454.

[29] Zhang, Z., Chen, H., Wu, W., Pang, W., Yan, G., Efficient removal of Alizarin Red S from aqueous solution by polyethyleneimine functionalized magnetic carbon nanotubes, Bioresource Technology, 2019, 293, 122100.

[30] Georgakilas, V., Otyepka, M., Bourlinos, A.B., Chandra, V., Kim, N., Kemp, K.C., et al., Functionalization of graphene: Covalent and non-covalent approaches, derivatives and applications, Chemical Reviews, 2012, 112, 6156–6214.

[31] Ferreira, F.V., Cividanes, L.D.S., Brito, F.S., Menezes, B.R.C.D., Franceschi, W., Simonetti, E.A.N., et al., Functionalization of carbon nanotube and applications, in Functionalizing Graphene and Carbon Nanotubes, Springer, 2016, 31–61.

[32] Sathyanarayana, S., Hübner, C., Thermoplastic nanocomposites with carbon nanotubes, in Structural Nanocomposites, Springer, 2013, 19–60.

[33] Hassan, T., Salam, A., Khan, A., Khan, S.U., Khanzada, H., Wasim, M., et al., Functional nanocomposites and their potential applications: A review, Journal of Polymer Research, 2021, 28, 1–22.

[34] Helbert, W., Cavaille, J., Dufresne, A., Thermoplastic nanocomposites filled with wheat straw cellulose whiskers. Part I: Processing and mechanical behavior, Polymer Composites, 1996, 17, 604–611.

[35] Ellis, T.S., D'Angelo, J.S., Thermal and mechanical properties of a polypropylene nanocomposite, Journal of Applied Polymer Science, 2003, 90, 1639–1647.

[36] Khlebtsov, B., Panfilova, E., Khanadeev, V., Bibikova, O., Terentyuk, G., Ivanov, A., et al., Nanocomposites containing silica-coated gold–silver nanocages and Yb-2, 4-Dimethoxyhematoporphyrin: Multifunctional capability of IR-luminescence detection, photosensitization, and photothermolysis, ACS Nano, 2011, 5, 7077–7089.

[37] Abdelrazek, E.M., Abdelghany, A.M., Badr, S.I., Morsi, M.A., Structural, optical, morphological and thermal properties of PEO/PVP blend containing different concentrations of biosynthesized Au nanoparticles, Journal of Materials Research and Technology, 2018, 7, 419–431.

[38] Makarchuk, O.V., Dontsova, T.A., Astrelin, I.M., Magnetic nanocomposites as efficient sorption materials for removing dyes from aqueous solutions, Nanoscale Research Letters, 2016, 11, 1–7.

[39] Ramesan, M., Jayakrishnan, P., Role of nickel oxide nanoparticles on magnetic, thermal and temperature dependent electrical conductivity of novel poly (vinyl cinnamate) based nanocomposites: Applicability of different conductivity models, Journal of Inorganic and Organometallic Polymers and Materials, 2017, 27, 143–153.

[40] De Moura, M.R., Mattoso, L.H., Zucolotto, V., Development of cellulose-based bactericidal nanocomposites containing silver nanoparticles and their use as active food packaging, Journal of Food Engineering, 2012, 109, 520–524.

[41] Arya, A., Sharma, A., Optimization of salt concentration and explanation of two peak percolation in blend solid polymer nanocomposite films, Journal of Solid State Electrochemistry, 2018, 22, 2725–2745.

[42] Hedayati, M.K., Faupel, F., Elbahri, M., Review of plasmonic nanocomposite metamaterial absorber, Materials, 2014, 7, 1221–1248.

[43] Volkert, A.A., Haes, A.J., Advancements in nanosensors using plastic antibodies, Analyst, 2014, 139, 21–31.

[44] Naka, K., Chujo, Y., Nanohybridized synthesis of metal nanoparticles and their organization, in Nanohybridization of Organic-inorganic Materials, Springer, 2009, 3–40.

[45] Porel, S., Ramakrishna, D., Hariprasad, E., Gupta, A.D., Radhakrishnan, T., Polymer thin film with in situ synthesized silver nanoparticles as a potent reusable bactericide, Current Science, 2011, 927–934.

[46] Raja, A., Ashokkumar, S., Marthandam, R.P., Jayachandiran, J., Khatiwada, C.P., Kaviyarasu, K., et al., Eco-friendly preparation of zinc oxide nanoparticles using Tabernaemontana divaricata and its photocatalytic and antimicrobial activity, Journal of Photochemistry and Photobiology B: Biology, 2018, 181, 53–58.

[47] Palza, H., Antimicrobial polymers with metal nanoparticles, International Journal of Molecular Sciences, 2015, 16, 2099–2116.

[48] Ravichandran, K., Nithiyadevi, K., Gobalakrishnan, S., Ganapathi Raman, R., Baneto, M., Swaminathan, K., et al., Enhancement of photocatalytic efficiency of ZnO nanopowders through Ag+ graphene addition, Materials Technology, 2016, 31, 865–871.

[49] Vella Durai, S., Ganapathi Raman, R., Kumar, E., Muthuraj, D., Structural, optical, morphological and thermal properties of CuO nanoparticles prepared by sol-gel technique, 2019.

[50] Le, T.-H., Kim, Y., Yoon, H., Electrical and electrochemical properties of conducting polymers, Polymers, 2017, 9, 150.

[51] Bergaya, F., Detellier, C., Lambert, J.-F., Lagaly, G., Introduction to clay–polymer nanocomposites (CPN), in Developments in Clay Science, Vol. 5, Elsevier, 2013, 655–677.

[52] Galimberti, M., Rubber-clay Nanocomposites: Science, Technology, and Applications, John Wiley & Sons, 2011.

[53] Şölener, M., Tunali, S., Özcan, A.S., Özcan, A., Gedikbey, T., Adsorption characteristics of lead (II) ions onto the clay/poly (methoxyethyl) acrylamide (PMEA) composite from aqueous solutions, Desalination, 2008, 223, 308–322.

[54] Rafiei, H., Shirvani, M., Ogunseitan, O., Removal of lead from aqueous solutions by a poly (acrylic acid)/bentonite nanocomposite, Applied Water Science, 2016, 6, 331–338.

[55] Setshedi, K.Z., Bhaumik, M., Songwane, S., Onyango, M.S., Maity, A., Exfoliated polypyrrole-organically modified montmorillonite clay nanocomposite as a potential adsorbent for Cr (VI) removal, Chemical Engineering Journal, 2013, 222, 186–197.

[56] Zare, Y., Shabani, I., Polymer/metal nanocomposites for biomedical applications, Materials Science and Engineering: C, 2016, 60, 195–203.

[57] Yashas, S.R., Shahmoradi, B., Wantala, K., Shivaraju, H.P., Potentiality of polymer nanocomposites for sustainable environmental applications: A review of recent advances, Polymer, 2021, 233, 124184.

[58] Duan, W., Li, M., Xiao, W., Wang, N., Niu, B., Zhou, L., et al., Enhanced adsorption of three fluoroquinolone antibiotics using polypyrrole functionalized Calotropis gigantea fiber, Colloids and Surfaces A: Physicochemical and Engineering Aspects, 2019, 574, 178–187.

[59] Zhu, Y., Peng, L., Fang, Z., Yan, C., Zhang, X., Yu, G., Structural engineering of 2D nanomaterials for energy storage and catalysis, Advanced Materials, 2018, 30, 1706347.

[60] Liu, W., Ullah, B., Kuo, -C.-C., Cai, X., Two-dimensional nanomaterials-based polymer composites: Fabrication and energy storage applications, Advances in Polymer Technology, 2019, 2019.

[61] Lin, Z., Goikolea, E., Balducci, A., Naoi, K., Taberna, P.-L., Salanne, M., et al., Materials for supercapacitors: When Li-ion battery power is not enough, Materials Today, 2018, 21, 419–436.

[62] Asen, P., Shahrokhian, S., Ternary nanostructures of Cr_2O_3/graphene oxide/conducting polymers for supercapacitor application, Journal of Electroanalytical Chemistry, 2018, 823, 505–516.

[63] Yang, C., Wei, H., Guan, L., Guo, J., Wang, Y., Yan, X., et al., Polymer nanocomposites for energy storage, energy saving, and anticorrosion, Journal of Materials Chemistry A, 2015, 3, 14929–14941.

[64] Shameem, M.M., Sasikanth, S., Annamalai, R., Raman, R.G., A brief review on polymer nanocomposites and its applications, Materials Today: Proceedings, 2021, 45, 2536–2539.

Elyor Berdimurodov*, Khasan Berdimuradov, Ilyos Eliboev,
Lazizbek Azimov, Yusufboy Rajabov, Jaykhun Mamatov,
Bakhtiyor Borikhonov, Abduvali Kholikov, Khamdam Akbarov

Chapter 3
Covalent and noncovalent surface functionalization of nanomaterials (for enhanced solubility, dispersibility and corrosion prevention potential)

Abstract: In the present times, the hybrids, smart hybrid, ceramic, organic–inorganic hybrid, carbon allotropes, polymeric, carbon dots (CDs) and heteroatom-doped CDs, graphene (G)/graphene oxide (GO), carbon nanotubes (SWCNTs/MWCNTs), quantum dots, mesoporous inorganic, MXenes, metal oxides, inorganic nanoparticles, silica and their derivatives mostly used surface-functionalized nanomaterials. The surface treatments of nanomaterials are important because various unique functions such as the hydrophobic performance, corrosion prevention potential, dispersibility, solubility and other various functional properties were enhanced. The surface functionalization was done by the covalent and noncovalent interactions. This modification depends on various factors such as the functional groups attached on the surface, surfactant nature, preparation methods, pH, solution temperature and other physical properties.

Keywords: Covalent and noncovalent functionalization, nanomaterials, surface treatments, nanotube, silica, corrosion protection, coatings

3.1 Introduction

In the present times, the nanomaterials were widely used in the corrosion protection. The surface treatments of nanomaterials are important because various unique functions such as the hydrophobic performance, corrosion prevention potential, dispersibility,

*Corresponding author: Elyor Berdimurodov, Faculty of Chemistry, National University of Uzbekistan, Tashkent 100034, Uzbekistan, e-mail: elyor170690@gmail.com
Khasan Berdimuradov, Faculty of Industrial Viticulture and Food Production Technology, Shahrisabz branch of Tashkent Institute of Chemical Technology, Shahrisabz 181306, Uzbekistan
Ilyos Eliboev, Lazizbek Azimov, Yusufboy Rajabov, Jaykhun Mamatov, Abduvali Kholikov,
Khamdam Akbarov, Faculty of Chemistry, National University of Uzbekistan, Tashkent 100034, Uzbekistan
Bakhtiyor Borikhonov, Faculty of Chemistry–Biology, Karshi State University, Karshi 130100, Uzbekistan

https://doi.org/10.1515/9783111071756-003

solubility and other various functional properties were enhanced [1–3]. The surface functionalization was done by the covalent and noncovalent interactions. This modification depends on the various factors such as the functional groups attached on the surface, surfactant nature, preparation methods, pH, solution temperature and other physical properties [4–8].

The smart, hybrid and organic–inorganic nanomaterials reveal various unique performances. In this action, the surface functionalization is important. The surface-modified materials are widely used in the corrosion protection, medical applications and pharmacy [9, 10]. In the functionalization processes, the special characters of the nanomaterials were developed. For example, the targeted surface character may be responsible for controlling the following factors: fluids composition, moisture, light, pH and temperature [11, 12].

The surface-functionalized nanomaterials were also named smart nanomaterials. The modern surface-functionalized nanomaterials were classified into the hybrid materials: inorganic (metalloid nanoparticles, metal oxide, metal and silica-based nanomaterials) and organic (mainly polymer-based nanoparticles) [4, 13, 14]. The functionalization of above materials developed their unique parameters. As a result, the new innovated nanomaterials are created. Especially, the corrosion protection performance of nanomaterials was developed by various covalent and noncovalent functionalization [3, 15, 16].

In the present times, the hybrids, smart hybrid, ceramic, organic–inorganic hybrid, carbon allotropes, polymeric, carbon dots (CDs) and heteroatom-doped CDs, graphene (G)/ graphene oxide (GO), carbon nanotubes (SWCNTs/MWCNTs), quantum dots, mesoporous inorganic, MXenes, metal oxides, inorganic nanoparticles, silica and their derivatives mostly used surface-functionalized nanomaterials [17–20]. The metallic materials were specially protected by using the abovementioned surface-functionalized nanomaterials. For example, Figure 3.1 shows the advanced surface functionalization of polypyrrole functionalized graphene oxide (GO-PPy): (a) pure epoxy coating, (b) GO-0.05% nanocomposite coating and (c) GP-0.05% nanocomposite coating [21]. In this functionalization, the GO was functionalized with the polypyrrole-based epoxy. As a result, the corrosion protection of pure epoxy was enhanced because of the synergism effect and surface modification of the nanomaterial and polymers. It is a covalent-type surface functionalization [22].

3.2 Covalent-surface functionalization of nanomaterials

The covalent interactions between the surface of nanomaterials and surfactant are important aspects. In corrosion protection, the nanomaterials were modified with the various surface modifications [23–25]. The aim of covalent-surface modification is to enhance solubility, dispersibility and corrosion prevention potential. For example, the various nanomaterials were covalent-modified with the organic compounds. For example, the

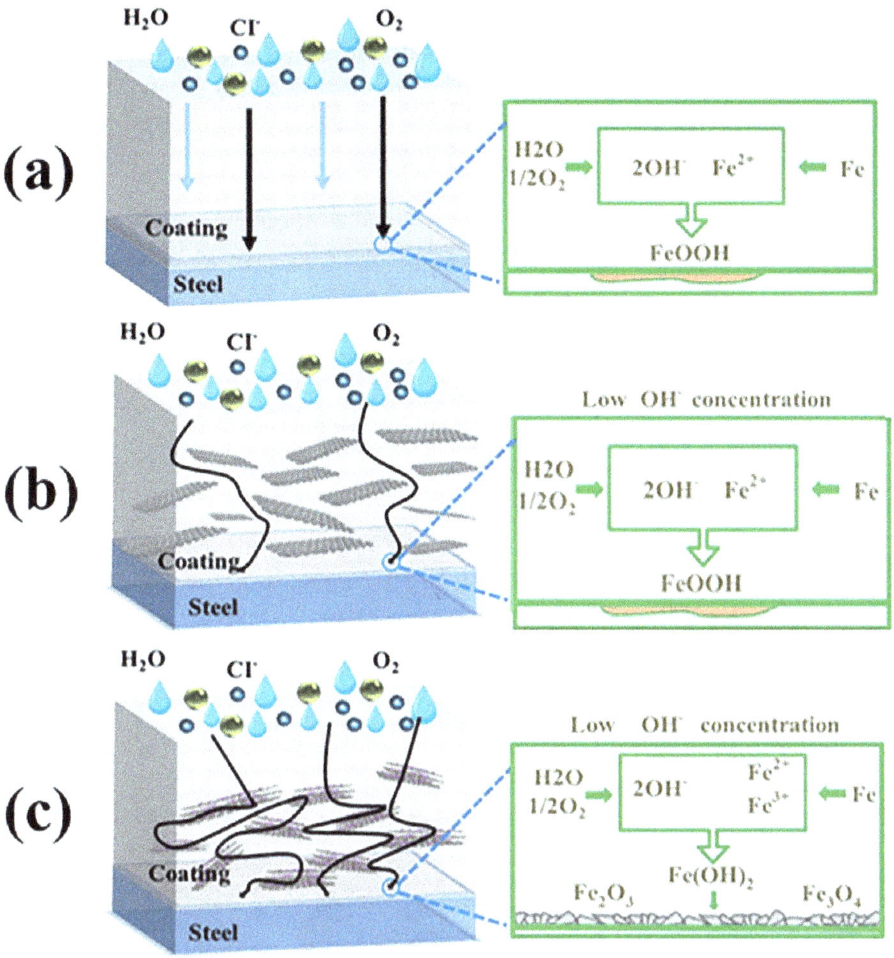

Figure 3.1: Advanced surface functionalization of polypyrrole functionalized graphene oxide (GO-PPy): (a) pure epoxy coating, (b) GO-0.05% nanocomposite coating and (c) GP-0.05% nanocomposite coating [21].

following main nanomaterials were used in the covalent-surface modification: the hybrids, smart hybrid, ceramic, organic–inorganic hybrid, carbon allotropes, polymeric, carbon dots (CDs) [16, 26] and heteroatom-doped CDs, graphene (G)/GO, carbon nanotubes (SWCNTs/MWCNTs), quantum dots, mesoporous inorganic, MXenes, metal oxides, inorganic nanoparticles, silica and their derivatives [27–29].

The silica-based nanomaterials were widely used in the corrosion protection. The reason for this is the silica surface was effectively modified with the various functional groups. As a result, the corrosion protection of these nanomaterials was enhanced. For example, Zhang et al. [30] modified the silica nanoparticles by the various covalent functionalizations. Figure 3.2 shows the covalent-surface modification of silica: SH-SiO$_2$

(Route A) and SH-SiO$_2$@BTA (Route B) [30]. The main aim of this surface modification in the silica nanomaterials is that the hydrophobic performance was enhanced. In this covalent interaction, the hydroxyl functional groups interacted with the surface of silica; as a result, the resin adhesive was attached to the silica surface. This attached resin adhesive is mainly responsible for the high hydrophobic performance.

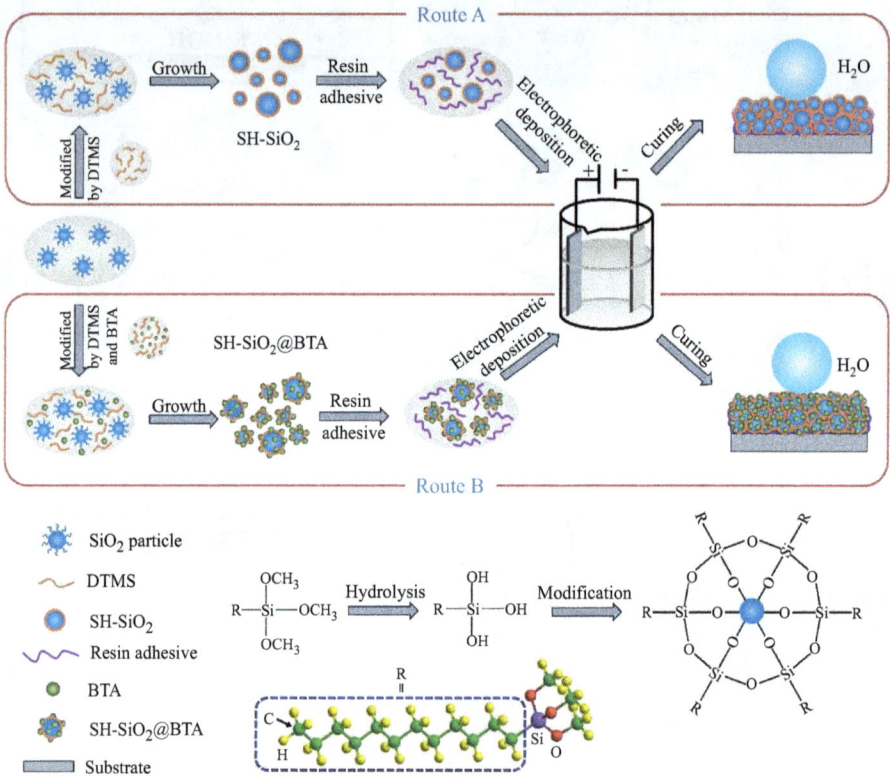

Figure 3.2: Covalent-surface modification of silica: SH-SiO$_2$ (Route A) and SH-SiO$_2$@BTA (Route B) [30].

Figure 3.3 shows the surface-covalent modification of silica nanoparticles: (a) surface modification with poly(ethylene glycol) and 2-mercaptobenzimidazol (MBI) [31]; (b) surface modification with stearic and oleic acid (γ-aminopropyltriethoxysilane) [32]. It is indicated that the nitrogen-contained imidazole compounds are good corrosion inhibitors. The amino functional groups are mainly responsible for the chemical and physical adsorption of corrosion protective agents. In the covalent interactions, Figure 3.3 represents the poly(ethylene glycol), 2-mercaptobenzimidazol and γ-aminopropyltriethoxysilane interacted efficiently with the surface of silica. As a result, the hydrophobic performance of silica was enhanced.

In the covalent modification, the organic inhibitors are attached to the surface of nanoparticles. The reason for this is the corrosion and viscoelastic properties were enhanced. The surface functionalization explored the structural analysis such as the FTIR, ^{1}H NMR and ^{13}C NMR analysis. The IR signals of molecules show the changes on the functional groups. The shifts in NMR signals of functional groups indicated the surface functionalization [33–38].

The covalent-surface modification effects on the change in the corrosion inhibition performance of nanoparticles. The modified nanomaterials significantly interacted with metal surface and adsorbed on the metal surface. The activation energy was enhanced with the adsorption of nanomaterials. The reason for this is that the corrosion processes required energy. The energy blocks were formed for the corrosion processes by the formation of protective layer on the metal surface [39–41].

The surface-covalent modification also effects on the charge transfer mechanism, in which the electron transfer between the anodic and cathodic regions was considerably depleted. The anodic metal dissolution released the free electrons while the hydrogen evolution required the free electrons. The rate in electron flows effects the corrosion and inhibition processes [42–44].

The next factor is that the inhibition mechanism of nanomaterials on the metal surface in the various corrosion mediums was effected by the covalent-surface modifications because the chemical structures of nanomaterials surface were changed. In the covalent-surface modification, the functional groups, aromatic rings and alkaline chains were changed. These changes are basic parameters in the corrosion-inhibition mechanism [45]. The covalent surface of silica chemically interacted with metal surface. The nanoparticles are large sized molecules; when they adsorbed on the metal surface, the large regions were insulated from the corrosion solution. The highest hydrophobic performance of nanomaterials is important to rise in the corrosion protection because the metal surface insulation depends on the values of hydrophobic performance [46, 47].

Zachariah et al. [48] investigated the corrosion protection performance of multiwalled carbon nanotubes for steel materials. The studied nanotube was surface modified with the polybenzoxazine. Figure 3.3 shows the surface-covalent modification of nanotube: (a) preparation of surface attached molecule (polybenzoxazine) and (b) covalent functionalization of nanotube [48]. It is indicated that the bisphenol A and oxydianiline (monomers) were polymerized to form the polybenzoxazine. In the next processes, the benzoyl peroxide was heated to form the free radicals, which was attached to the polymer chains. The aim of these free radicals is that they are responsible for the linking bridges between the polymer chain and nanotube. Finally, the radical-contained polymer chain was attached to carbon nanotube to form the protective nanoparticles. The dispersion homogeneity of nanotube was enhanced by the matrix-polymer functionalization on the surface of nanotube. The protective performance of coating depends on the dispersion homogeneity. It was found that the corrosion protection was over 99.91% after the surface-covalent modification. The corrosion rate was 20 µm/year while this value was decreased to 0.179 µm/year with the addition surface-modified nanotube.

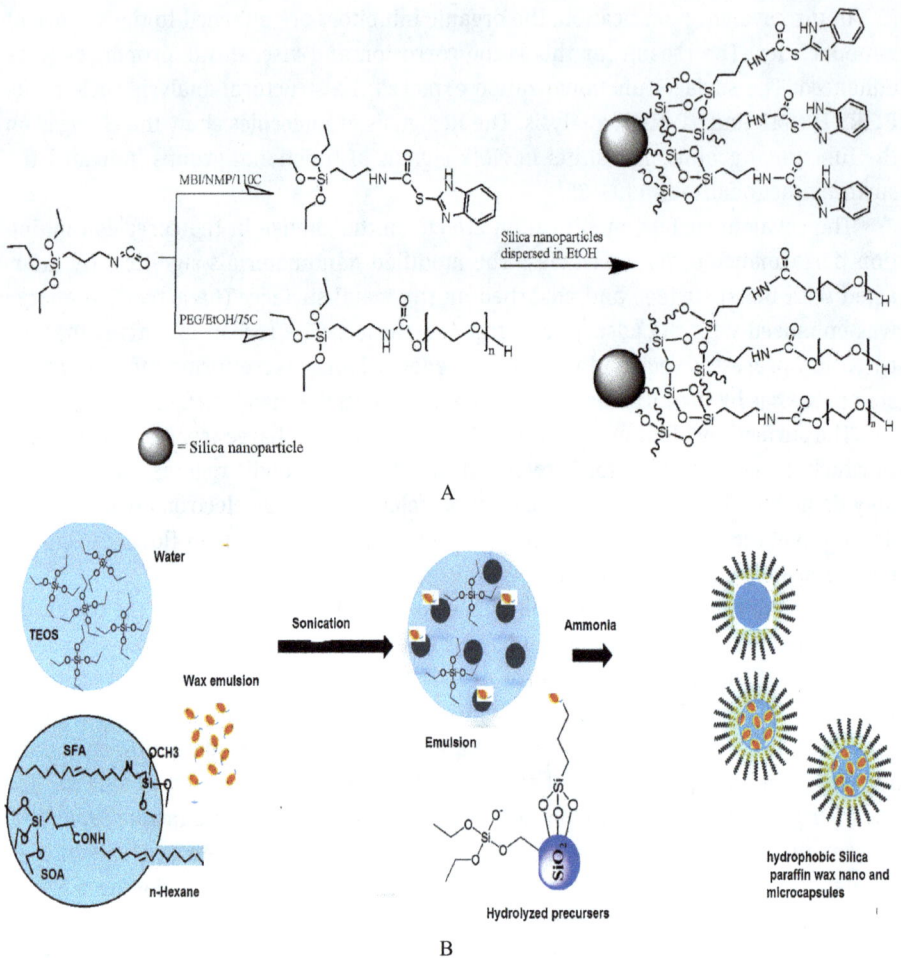

Figure 3.3: Surface-covalent modification of silica nanoparticles: (a) surface modification with poly (ethylene glycol) and 2-mercaptobenzimidazol (MBI) [31]; (b) surface modification with stearic and oleic acid (γ-aminopropyltriethoxysilane) [32].

These results confirmed the effects of surface-covalent modification in enhancing the corrosion protection for the carbon-based nanotube.

Shirazi et al. [49] studied the anticorrosion properties of carbon nanotube and graphene nanomaterials for steel. These nanomaterials were functionalized with the poly(o-anthranilic acid) by the surface-covalent interactions. Their corrosion performance that was checked in the 2 M hydrochloric acid solution shows that the metal surface is efficiently protected (Figure 3.4). The purpose of this covalent-surface modification is that the nitrogen-contained corrosion inhibitor was attached on the surface of nanomaterials. The corrosion inhibitor efficiently linked with the metal surface by

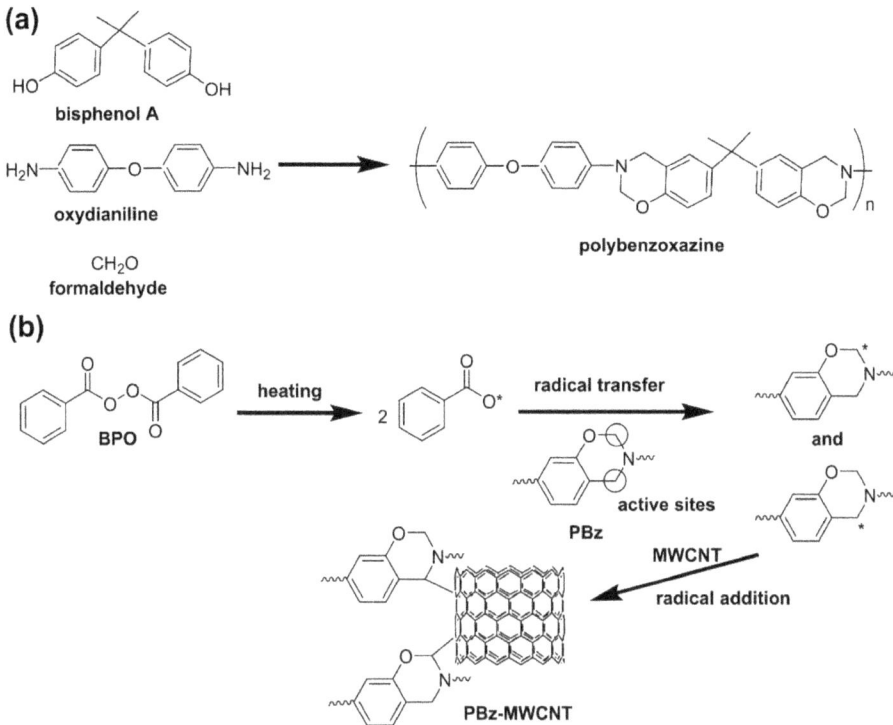

Figure 3.4: Surface-covalent modification of nanotube: (a) preparation of surface attached molecule (polybenzoxazine); (b) covalent functionalization of nanotube [48].

the formation of covalent interactions. As a result, the nanotube covered metal surface efficiently. The formed protective layer on the metal surface is more stable in the 2 M hydrochloric acid solution. The corrosion solution is more aggressive and stability of protective layer in the acidic solution is a very important factor [50–52].

3.3 Noncovalent-surface functionalization of nanomaterials

The polymer-based protective coatings are vastly used in the metal protection. However, the protective nanomaterials are more UV-sensitive. This sensitivity is the reason for the destruction of nanomaterials on the metal surface. This UV sensitivity was blocked with the noncovalent modification of nanomaterials. The corrosion inhibition was slowly decreased with the higher UV photodegradation. In the corrosion science, the inorganic nanocarriers were mostly used as an addition in the protective nanomaterials. The inorganic compounds interacted well with the surface of nanomaterials.

This interaction depends on the donor-acceptor performances of the surface of nano-materials. Figure 3.5 shows the surface-noncovalent functionalizing by copper: (a) change in reaction rate of benzotriazole (BTA) with the surface of mesoporous silica nanoparticles (MSN); (b) BTA solution and (c) interaction between copper and BTA; (d) structures of obtained MSN [53]. In this noncovalent interaction, the copper was reacted with the BTA to form the Cu-BTA complex, which was attached on the surface of silica mesoporous. The UV attacks on the coating surface were blocked with the functionalization of Cu-BTA complex.

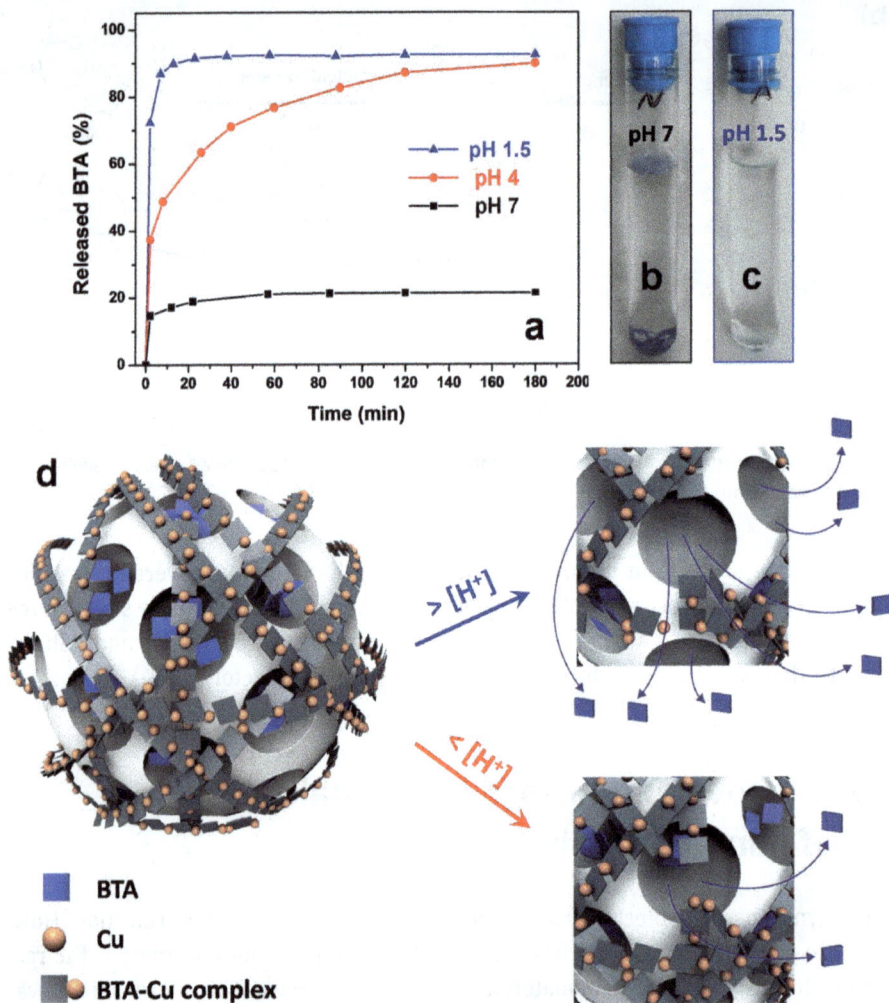

Figure 3.5: Surface-noncovalent functionalizing by copper: (a) change in reaction rate of benzotriazole (BTA) with the surface of mesoporous silica nanoparticles (MSN); (b) BTA solution and (c) interaction between copper and BTA; (d) structures of obtained MSN [53].

Nguyen et al. [54] explored the cerium (Ce)-silica modification and studied its corrosion protection performance for steel. In this surface-noncovalent interaction, the free electrons interacted with the d-orbitals of Ce. The surface-noncovalent interactions depend on the d-orbital system of Ce. In this action, the Ce is acceptor centers and the oxygen atoms are donor regions. The electrons are shared from the donor to acceptor. As a result, the noncovalent bonds between the silica surface and Ce ions were formed. The surface noncovalent of nanomaterials can enhance the corrosion protection performance.

Figure 3.6 shows the noncovalent-surface modification: (a) preparation of Zn/polyaniline/silica nanoparticles [55]; (b) surface modification of mesoporous silica nanoparticles with Eriochrom Black T. [56]. It is clear from the illustrations that the aniline contained NH functional groups, which have unpaired electrons. The zinc ions have more vacant d-orbitals. In the noncovalent interactions, the unpaired electrons of nitrogen are shared to the vacant d-orbitals of zinc. In comparison, the nitrate ions are also electrostatically linked with the NH functional groups. These electrostatically interactions are the main reason for the dipole-dipole interactions on the surface of nanomaterials. The surface of nanomaterials was built with the aniline polymers. The metals and anions easily functionalized with the surface of nanomaterials by the electrostatic or dipole-dipole interactions. The main reason for the zinc ions based the noncovalent-surface modification to develop in the chemical interaction, dispersion, thermal and rheological properties. The metal protection and corrosion barrier performance of paint-based coating was enhanced with the zinc functionalizing.

It is also revealed in Figure 3.6 that the surface of nanomaterials was modified with Eriochrom Black T (ECBT) by the noncovalent-surface modification. In this process, first, the amino functional groups were attached to silica nanoparticle. Next, the iron ions were attached to amino functional group by the donor-acceptor mechanism. As a result, the Fe=N bonds were formed. In the third step, the Eriochrom Black T electrostatically interacted with the nanoparticle surface by the electrostatic interaction between the metal and sulfo functional groups. The main reasons in the noncovalent interactions are that

(i)   Eriochrom Black T was a good corrosion inhibitor. When it was attached to the surface of nanomaterials, its corrosion protection performance was enhanced.

(ii) Self-healing process is enhanced with the functionalization of Eriochrom Black T, means that when the metal surface was destructed or corroded, the corrosion inhibitors are released from the nanomaterials matrix. As a result, the metal surface was inhibited with the released corrosion inhibitors.

(iii) The metal in 3.5 wt% NaCl solution was effectively protected by the surface-modified nanoparticles.

In other research work, Najmi et al. [57] prepared the modified nanotube with the zinc and polyaniline. In this modification, the nanotube was oxidized and coated with polyaniline. Finally, the formed nanomaterials are mixed with the zinc salts at the

(A)

(B)

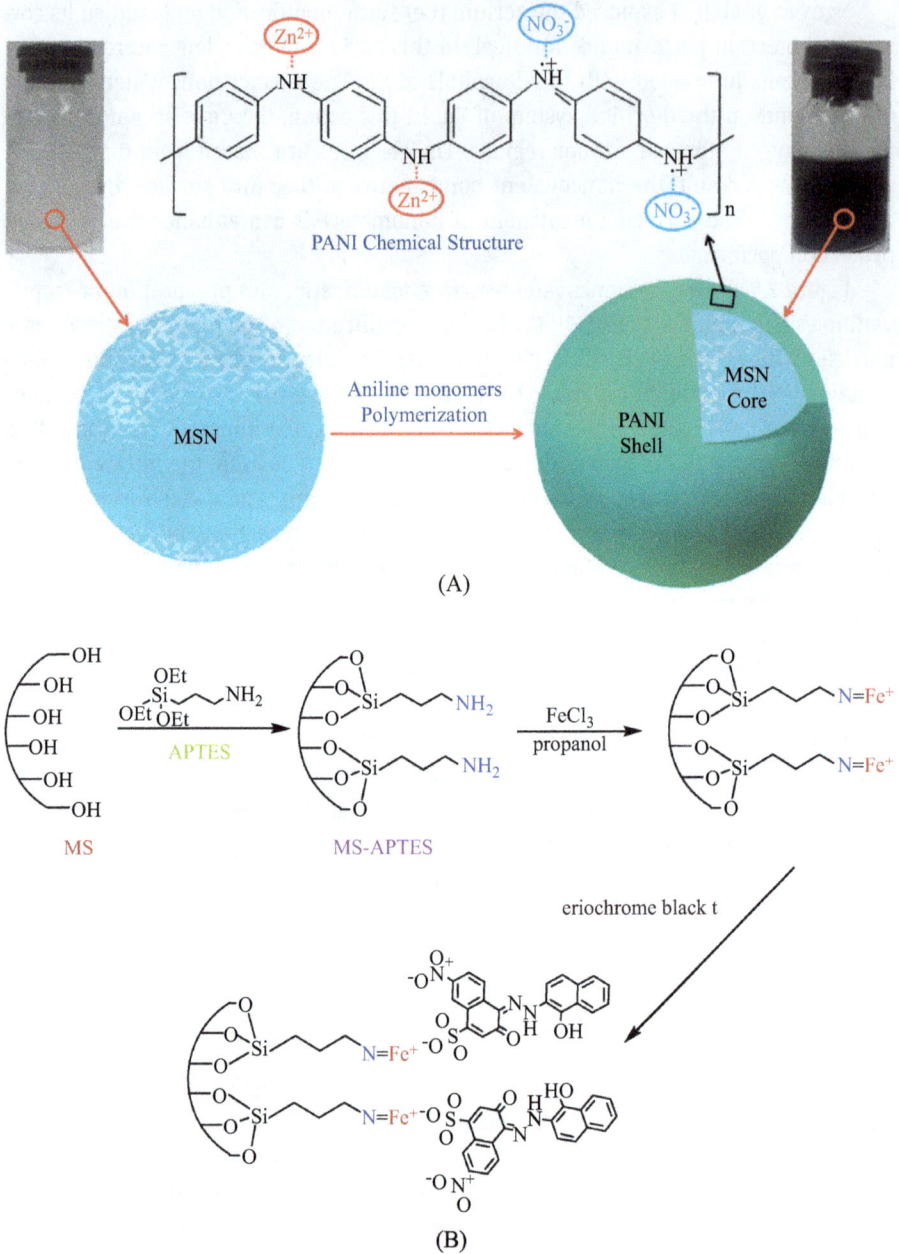

Figure 3.6: Noncovalent-surface modification: (a) preparation of Zn/polyaniline/silica nanoparticles [55]; and (b) surface modification of mesoporous silica nanoparticles with Eriochrom Black T [56].

alkaline solution. Then the nanoparticle was functionalized with zinc atoms by the surface-noncovalent interactions.

The surface regions, surface nature, number of functional groups, environmental factors and surfactant structures are basic factors in the surface-noncovalent interactions. The nanotubes have large surface area, and it is more convenient to surface modification [58–62].

The Ni metal was used in the surface-noncovalent modification of nanotube [63], thus functionalizing to enhance the corrosion performance of metallic materials. This modified nanotube can protect the steel pipes from the high flow rate.

3.4 Future perspectives in the covalent and noncovalent-surface functionalization of nanomaterials

The surface-covalent modification also effects on the charge transfer mechanism, in which the electron transfer between the anodic and cathodic regions was considerably depleted. The anodic metal dissolution releases the free electrons while the hydrogen evolution requires the free electrons. The rate in electron flows effects the corrosion and inhibition processes [64–68].

The noncovalent interactions on the surface of nanoparticles influence the corrosion protection, self-healing and other basic properties. The nanocarriers are the main part of nanomaterials in the corrosion protection; this part contained the corrosion inhibitors. When the metal surface was protected, the corrosion inhibitors are released from the carriers to the metal surface [69].

In noncovalent interactions on the metal surface, the metal ions are mainly responsible for the dipole–dipole relationship. There are two types of noncovalent-surface modification:

(i) In the first modification, the metal ions were attached on to the surface of nanomaterials. The aim of this metal modification is that the corrosion protection, hydrophobic performance and anti-UV properties were enhanced [70, 71].

(ii) In the second functionalization, the metal ions have bridge role. It means that the some corrosion inhibitors are attached on to the surface of protective nanomaterials by metal bridges. For example, the nitrogen-contained corrosion inhibitors are more efficient. The surface modification of nanomaterials with the corrosion inhibitor is difficult. Then the metal ions were used in the corrosion inhibitor modification on the surface of nanomaterials. Then, the first, metal ions were interacted with the nanomaterials surface by the noncovalent interactions. In this case, the metal ions are charged positive while the surface of nanomaterials was charged negatively. As a result, the positively charged metal ions are easily interconnected with the surface of nanomaterials. In the second stage, the corrosion inhibitors are linked to the surface of nanomaterials by the metal ions. The attached corrosion inhibitor promotes the corrosion performance of nanomaterials [72, 73].

Additionally, the pH sensitivity, self-repairing and lifetime of corrosion-protective nano-materials were enhanced with the noncovalent-surface modification. Some nanomaterials are slowly destructed in the rise of pH. In this case, the surface of nanomaterials and the pH resistance compounds will be modified by the covalent and noncovalent-surface functionalization of nanomaterials. The self-repairing and lifetime of corrosion protective nanomaterials will be enhanced by the covalent and noncovalent-surface functionalization of nanomaterials. The surface-modified nanomaterials will dominate in the corrosion protection applications [19].

3.5 Conclusion

In the present times, the hybrids, smart hybrid, ceramic, organic–inorganic hybrid, carbon allotropes, polymeric, carbon dots (CDs) and heteroatom-doped CDs, graphene (G)/GO, carbon nanotubes (SWCNTs/MWCNTs), quantum dots, mesoporous inorganic, MXenes, metal oxides, inorganic nanoparticles, silica and their derivatives mostly used surface-functionalized nanomaterials. The surface treatments of nanomaterials are important because the various unique functions such as the hydrophobic performance, corrosion prevention potential, dispersibility, solubility and other various functional properties were enhanced. The surface functionalization was done by the covalent and noncovalent interactions. This modification depends on the various factors such as the functional groups attached on the surface, surfactant nature, preparation methods, pH, solution temperature and other physical properties. The surface-covalent modification also effects on the charge transfer mechanism, in which the electron transfer between the anodic and cathodic regions was considerably depleted. The anodic metal dissolution releases the free electrons while the hydrogen evolution required the free electrons. The rate in electron flows effects the corrosion and inhibition processes. The noncovalent interactions on the surface of nanoparticles influence the corrosion protection, self-healing and other basic properties. The nanocarriers are main part of nanomaterials in the corrosion protection; this part contained the corrosion inhibitors. When the metal surface was protected, the corrosion inhibitors are released from the carriers to the metal surface.

References

[1]   Yan, D., et al., Dual-functional graphene oxide-based nanomaterial for enhancing the passive and active corrosion protection of epoxy coating, Composites Part B Engineering, 2021, 222, 109075.

[2]   Vitharana, M.G., et al., A study on strength and corrosion protection of cement mortar with the inclusion of nanomaterials, Sustainable Materials and Technologies, 2020, 25, e00192.

[3]   AhadiParsa, M., et al., Rising of MXenes: Novel 2D-functionalized nanomaterials as a new milestone in corrosion science-a critical review, Advances in Colloid and Interface Science, 2022, 102730.

[4]   Pourhashem, S., et al., Polymer/Inorganic nanocomposite coatings with superior corrosion protection performance: A review, Journal of Industrial and Engineering Chemistry, 2020, 88, 29–57.

[5]   Khodair, Z.T., Khadom, A.A., Jasim, H.A., Corrosion protection of mild steel in different aqueous media via epoxy/nanomaterial coating: Preparation, characterization and mathematical views, Journal of Materials Research and Technology, 2019, 8(1), 424–435.

[6]   Zhu, M., et al., Insights into the newly synthesized N-doped carbon dots for Q235 steel corrosion retardation in acidizing media: A detailed multidimensional study, Journal of Colloid and Interface Science, 2022, 608, 2039–2049.

[7]   Verma, D.K., et al., N-hydroxybenzothioamide derivatives as green and efficient corrosion inhibitors for mild steel: Experimental, DFT and MC simulation approach, Journal of Molecular Structure, 2021, 1241, 130648.

[8]   Verma, D.K., et al., Ionic liquids as green and smart lubricant application: An overview, Ionics, 2022, 1–10.

[9]   Yan, D., et al., Smart self-healing coating based on the highly dispersed silica/carbon nanotube nanomaterial for corrosion protection of steel, Progress in Organic Coatings, 2022, 164, 106694.

[10]   Mohammadpour, Z., Zare, H.R., Structural effect of different carbon nanomaterials on the corrosion protection of Ni–W alloy coatings in saline media, New Journal of Chemistry, 2018, 42(7), 5425–5432.

[11]   Wieszczycka, K., et al., Surface functionalization–The way for advanced applications of smart materials, Coordination Chemistry Reviews, 2021, 436, 213846.

[12]   Dave, P.N., Chopda, L.V., Sahu, L., Applications of nanomaterials in corrosion protection inhibitors and coatings, in Functionalized Nanomaterials for Corrosion Mitigation: Synthesis, Characterization, and Applications, ACS Publications, 2022, 189–212.

[13]   Chukwuike, V.I., et al., Capped and uncapped nickel tungstate (NiWO4) nanomaterials: A comparison study for anti-corrosion of copper metal in NaCl solution, Corrosion Science, 2019, 158, 108101.

[14]   Duran, B., Pat, S., Improved corrosion protection of stainless steel by two dimensional BN nanomaterial coating, ECS Journal of Solid State Science and Technology, 2022, 11(6), 063017.

[15]   Liu, C., et al., Synthesis of l-histidine-attached graphene nanomaterials and their application for steel protection, ACS Applied Nano Materials, 2018, 1(3), 1385–1395.

[16]   Shahmoradi, A.R., et al., Theoretical and surface/electrochemical investigations of walnut fruit green husk extract as effective inhibitor for mild-steel corrosion in 1M HCl electrolyte, Journal of Molecular Liquids, 2021, 338, 116550.

[17]   Peng, T., et al., Polymer nanocomposite-based coatings for corrosion protection, Chemistry–An Asian Journal, 2020, 15(23), 3915–3941.

[18]   Sharma, S., Kumar, A., MXenes and MXene-based nanomaterials for corrosion protection, Materials Letters, 2022, 133789.

[19]   Aslam, R., Mobin, M., Aslam, J., Nanomaterials as corrosion inhibitors, in Inorganic Anticorrosive Materials, Elsevier, 2022, 3–20.

[20]   Al-Akhras, N., Makableh, Y., Dagamseh, D.A., Evaluating composite nanomaterials to control corrosion of reinforcing steel using different tests, Case Studies in Construction Materials, 2022, 16, e00963.

[21] Zhu, Q., et al., Epoxy coating with in-situ synthesis of polypyrrole functionalized graphene oxide for enhanced anticorrosive performance, Progress in Organic Coatings, 2020, 140, 105488.

[22] Verma, C., Quraishi, M.A., Carbohydrate polymers-modified carbon allotropes for enhanced anticorrosive activity: State-of-arts and perspective, Chemical Engineering Journal Advances, 2022, 100428.

[23] Nazari, M.H., et al., Nanocomposite organic coatings for corrosion protection of metals: A review of recent advances, Progress in Organic Coatings, 2022, 162, 106573.

[24] Yeganeh, M., et al., Enhancement routes of corrosion resistance in the steel reinforced concrete by using nanomaterials, in Smart Nanoconcretes and Cement-Based Materials, Elsevier, 2020, 583–599.

[25] Jiang, F., et al., A polyethyleneimine-grafted graphene oxide hybrid nanomaterial: Synthesis and anti-corrosion applications, Applied Surface Science, 2019, 479, 963–973.

[26] Berdimurodov, E., et al., The recent development of carbon dots as powerful green corrosion inhibitors: A prospective review, Journal of Molecular Liquids, 2021, 118124.

[27] Kavimani, V., et al., Corrosion protection behaviour of r-GO/TiO2 hybrid composite coating on magnesium substrate in 3.5 wt.% NaCl, Progress in Organic Coatings, 2018, 125, 358–364.

[28] Mostafa, S.A., et al., Evaluation of the nano silica and nano waste materials on the corrosion protection of high strength steel embedded in ultra-high performance concrete, Scientific Reports, 2021, 11(1), 1–16.

[29] Ye, X., et al., In-situ growth of NiAl-layered double hydroxide on AZ31 Mg alloy towards enhanced corrosion protection, Nanomaterials, 2018, 8(6), 411.

[30] Zhang, X.-F., et al., Robust superhydrophobic coatings prepared by cathodic electrophoresis of hydrophobic silica nanoparticles with the cationic resin as the adhesive for corrosion protection, Corrosion Science, 2020, 173, 108797.

[31] Khodabakhshi, J., Mahdavi, H., Najafi, F., Investigation of viscoelastic and active corrosion protection properties of inhibitor modified silica nanoparticles/epoxy nanocomposite coatings on carbon steel, Corrosion Science, 2019, 147, 128–140.

[32] Atta, A.M., et al., New hydrophobic silica nanoparticles capped with petroleum paraffin wax embedded in epoxy networks as multifunctional steel epoxy coatings, Progress in Organic Coatings, 2019, 128, 99–111.

[33] Ramezanzadeh, B., et al., Synthesis and characterization of polyaniline tailored graphene oxide quantum dot as an advance and highly crystalline carbon-based luminescent nanomaterial for fabrication of an effective anti-corrosion epoxy system on mild steel, Journal of the Taiwan Institute of Chemical Engineers, 2019, 95, 369–382.

[34] Tavandashti, N.P., Almas, S.M., Esmaeilzadeh, E., Corrosion protection performance of epoxy coating containing alumina/PANI nanoparticles doped with cerium nitrate inhibitor on Al-2024 substrates, Progress in Organic Coatings, 2021, 152, 106133.

[35] Bouibed, A., Doufnoune, R., Synthesis and characterization of hybrid materials based on graphene oxide and silica nanoparticles and their effect on the corrosion protection properties of epoxy resin coatings, Journal of Adhesion Science and Technology, 2019, 33(8), 834–860.

[36] Rbaa, M., et al., Synthesis of new halogenated compounds based on 8-hydroxyquinoline derivatives for the inhibition of acid corrosion: Theoretical and experimental investigations, Materials Today Communications, 2022, 33, 104654.

[37] Rbaa, M., et al., Development process for eco-friendly corrosion inhibitors, in Eco-Friendly Corrosion Inhibitors, Elsevier, 2022, 27–42.

[38] Berdimurodov, E., et al., Green β-cyclodextrin-based corrosion inhibitors: Recent developments, innovations and future opportunities, Carbohydrate Polymers, 2022, 119719.

[39] Ashraf, M.A., et al., Amino acid and TiO2 nanoparticles mixture inserted into sol-gel coatings: An efficient corrosion protection system for AZ91 magnesium alloy, Progress in Organic Coatings, 2019, 136, 105296.

[40] Chen, H., et al., Highly hydrophobic polyaniline nanoparticles for anti-corrosion epoxy coatings, Chemical Engineering Journal, 2021, 420, 130540.

[41] Berdimurodov, E., et al., MOFs-based corrosion inhibitors, in Supramolecular Chemistry in Corrosion and Biofouling Protection, CRC Press, 2021, 287–305.

[42] Ashassi-Sorkhabi, H., Moradi-Alavian, S., Kazempour, A., Salt-nanoparticle systems incorporated into sol-gel coatings for corrosion protection of AZ91 magnesium alloy, Progress in Organic Coatings, 2019, 135, 475–482.

[43] Samadianfard, R., et al., Oxidized fullerene/sol-gel nanocomposite for corrosion protection of AM60B magnesium alloy, Surface and Coatings Technology, 2020, 385, 125400.

[44] Berdimurodov, E., et al., Inhibition properties of 4,5-dihydroxy-4,5-di-*p*-tolylimidazolidine-2-thione for use on carbon steel in an aggressive alkaline medium with chloride ions: Thermodynamic, electrochemical, surface and theoretical analyses, Journal of Molecular Liquids, 2021, 327, 114813.

[45] Berdimurodov, E., et al., Novel glycoluril pharmaceutically active compound as a green corrosion inhibitor for the oil and gas industry, Journal of Electroanalytical Chemistry, 2022, 907, 116055.

[46] Zhang, F., et al., The effect of functional graphene oxide nanoparticles on corrosion resistance of waterborne polyurethane, Colloids and Surfaces A: Physicochemical and Engineering Aspects, 2020, 591, 124565.

[47] Dippong, T., Levei, E.A., Cadar, O., Recent advances in synthesis and applications of MFe2O4 (M=Co, Cu, Mn, Ni, Zn) nanoparticles, Nanomaterials, 2021, 11(6), 1560.

[48] Zachariah, S., Liu, Y.-L., Nanocomposites of polybenzoxazine-functionalized multiwalled carbon nanotubes and polybenzoxazine for anticorrosion application, Composites Science and Technology, 2020, 194, 108169.

[49] Shirazi, Z., Golikand, A.N., Keshavarz, M.H., A new nanocomposite based on poly (o-anthranilic acid), graphene oxide and functionalized carbon nanotube as an efficient corrosion inhibitor for stainless steel in severe environmental corrosion, Composites Communications, 2020, 22, 100467.

[50] Nayak, S.R., et al., Functionalized multi-walled carbon nanotube/polyindole incorporated epoxy: An effective anti-corrosion coating material for mild steel, Journal of Alloys and Compounds, 2021, 856, 158057.

[51] Berdimurodov, E., et al., Experimental and theoretical assessment of new and eco–friendly thioglycoluril derivative as an effective corrosion inhibitor of St2 steel in the aggressive hydrochloric acid with sulfate ions, Journal of Molecular Liquids, 2021, 335, 116168.

[52] Berdimurodov, E., et al., Novel gossypol–indole modification as a green corrosion inhibitor for low–carbon steel in aggressive alkaline–saline solution, Colloids and Surfaces A: Physicochemical and Engineering Aspects, 2022, 637, 128207.

[53] Castaldo, R., et al., On the acid-responsive release of benzotriazole from engineered mesoporous silica nanoparticles for corrosion protection of metal surfaces, Journal of Cultural Heritage, 2020, 44, 317–324.

[54] Nguyen, T.V., et al., Ce-loaded silica nanoparticles in the epoxy nanocomposite coating for anticorrosion protection of carbon steel, Anti-Corrosion Methods and Materials, 2022, 69(5), 514–523.

[55] Haddadi, S.A., et al., Zinc-doped silica/polyaniline core/shell nanoparticles towards corrosion protection epoxy nanocomposite coatings, Composites Part B Engineering, 2021, 212, 108713.

[56] Ashrafi-Shahri, S.M., Ravari, F., Seifzadeh, D., Smart organic/inorganic sol-gel nanocomposite containing functionalized mesoporous silica for corrosion protection, Progress in Organic Coatings, 2019, 133, 44–54.

[57] Najmi, P., et al., Synthesis and application of Zn-doped polyaniline modified multi-walled carbon nanotubes as stimuli-responsive nanocarrier in the epoxy matrix for achieving excellent barrier-self-healing corrosion protection potency, Chemical Engineering Journal, 2021, 412, 128637.

[58] Kaur, J., et al., Euphorbia prostrata as an eco-friendly corrosion inhibitor for steel: Electrochemical and DFT studies, Chemical Papers, 2022, 1–20.

[59] Haldhar, R., et al., Corrosion inhibitors: Industrial applications and commercialization, in Sustainable Corrosion Inhibitors II: Synthesis, Design, and Practical Applications, American Chemical Society, 2021, 10–219.

[60] Dewangan, Y., et al., N-hydroxypyrazine-2-carboxamide as a new and green corrosion inhibitor for mild steel in acidic medium: Experimental, surface morphological and theoretical approach, Journal of Adhesion Science and Technology, 2022, 1–21.

[61] Dagdag, O., et al., Rheological and simulation for macromolecular matrix epoxy bi-functional aromatic amines, Polymer Bulletin, 2021.

[62] Berdimurodov, E., et al., A gossypol derivative as an efficient corrosion inhibitor for St2 steel in 1 M HCl + 1 M KCl: An experimental and theoretical investigation, Journal of Molecular Liquids, 2021, 328, 115475.

[63] Prasannakumar, R.S., et al., Electrochemical and hydrodynamic flow characterization of corrosion protection persistence of nickel/multiwalled carbon nanotubes composite coating, Applied Surface Science, 2020, 507, 145073.

[64] Zhao, W., et al., Environmentally-friendly superhydrophobic surface based on Al2O3@ KH560@ SiO2 electrokinetic nanoparticle for long-term anti-corrosion in sea water, Applied Surface Science, 2019, 484, 307–316.

[65] Dagdag, O., et al., Synthesis, physicochemical properties, theoretical and electrochemical studies of Tetraglycidyl methylenedianiline, Journal of Molecular Structure, 2022, 133508.

[66] Dagdag, O., et al., Recent progress in epoxy resins as corrosion inhibitors: Design and performance, Journal of Adhesion Science and Technology, 2022, 1–22.

[67] Dagdag, O., et al., Graphene and graphene oxide as nanostructured corrosion inhibitors., Carbon Allotropes: Nanostructured Anti-Corrosive Materials, 2022, 133.

[68] Berdimurodov, E., et al., Novel bromide–cucurbit[7]uril supramolecular ionic liquid as a green corrosion inhibitor for the oil and gas industry, Journal of Electroanalytical Chemistry, 2021, 901, 115794.

[69] Kaseem, M., Ko, Y.G., A novel hybrid composite composed of albumin, WO3, and LDHs film for smart corrosion protection of Mg alloy, Composites Part B: Engineering, 2021, 204, 108490.

[70] Dagdag, O., et al., Functionalized nanomaterials for corrosion mitigation: Synthesis, characterization & applications, in Functionalized Nanomaterials for Corrosion Mitigation: Synthesis, Characterization, and Applications, ACS Publications, 2022, 67–85.

[71] Dagdag, O., et al., Epoxy coating as effective anti-corrosive polymeric material for aluminum alloys: Formulation, electrochemical and computational approaches, Journal of Molecular Liquids, 2021, 117886.

[72] Bahgat Radwan, A., et al., Electrospun highly corrosion-resistant polystyrene–nickel oxide superhydrophobic nanocomposite coating, Journal of Applied Electrochemistry, 2021, 51, 1605–1618.

[73] Berdimurodov, E., et al., Novel cucurbit[6]uril-based [3]rotaxane supramolecular ionic liquid as a green and excellent corrosion inhibitor for the chemical industry, Colloids and Surfaces A: Physicochemical and Engineering Aspects, 2022, 633, 127837.

Aysun Aksu, Hüseyin Fatih Çetinkaya, Serap Çetinkaya*,
Gamze Tüzün, Burak Tüzün

Chapter 4
MXenes and their composites as corrosion prevention

Abstract: MXene, an advanced metal-based 2D ceramic material family, has had a significant impact in the field of corrosion prevention due to its unique physicochemical properties. Due to the need for further advancement and improvement in MXene materials, studies have turned to the manufacture of composites that help strengthen MXene composites in terms of their properties and applications in various fields. MXenes have excellent filling properties. Anticorrosive application of MXene-filled polymer coatings has been extensively reported. Although modified graphene oxide is widely used as a filler in boron nitride and epoxy-based coatings, its applications as filler in other polymers with new properties are needed. This chapter focuses on MXenes as corrosion inhibitors, the properties of their composites and their application in industry.

Keywords: Corrosion, corrosion prevention, metals, MXenes

4.1 Introduction

Corrosion is an industrial challenge in the survival performance of metal-based hardware. A solution to this problem has been to develop corrosion-resistant surfaces [1, 2] and to this end one of the outstanding approaches has been the surface coating especially with organic compounds [3, 4].

Currently, one can mention about an invaluable inventory of anticorrosive substances including epoxy, aliphatic and aromatic matrix materials. They are known for their spatial and structural stabilities. They exert excellent insulation capacities in highly adverse environments [5]. Here a newly emerging shortcoming has been the

*Corresponding author: Serap Çetinkaya,** Department of Molecular Biology and Genetics, Science Faculty, Sivas Cumhuriyet University, Sivas 58140, Turkey, e-mail: serapcetinkaya2012@gmail.com
Aysun Aksu, Department of Molecular Biology and Genetics, Science Faculty, Sivas Cumhuriyet University, Sivas 58140, Turkey
Hüseyin Fatih Çetinkaya, Department of Environmental Engineering, Faculty of Engineering, Sivas Cumhuriyet University, Sivas, Turkey
Gamze Tüzün, Department of Chemistry, Faculty of Science, Sivas Cumhuriyet University, Sivas, Turkey
Burak Tüzün, Plant and Animal Production Department, Technical Sciences Vocational School of Sivas, Sivas Cumhuriyet University, Sivas, Turkey

https://doi.org/10.1515/9783111071756-004

formation of micropores within the coating composite. Such deformations have been challenged by the addition of suitable crosslinkers [4, 6].

The anticorrosive material has to be both harmless to the environment and relatively cost-effective. Some examples of these include graphene oxide (GO), boron nitride (h-BN) and epoxy-modified GO, which are the good substitutes of heavy metals [7, 8].

Another alternative, at present, is the electrical and thermal conductor MXene combining both ceramic and steel characteristics [9]. It has the formula $M_{n+1}AX_n$: M, an early transition metal; A, one of the elements of IIIA or IV A group; X, carbon or nitrogen; and $n = 1$, 2, or 3. M and A are held with a metallic-, covalent- or ionic bond, while $M - X$ is formed covalently [10]. The A layer interlocks the M layers inside the MAX and the forms octahedral structures in which X atoms reside [11]. This compound, harboring differing bond energies, enables selective etching processes to be implemented and hence form MXene [11, 12]. In the etching process, the surface of the $M_{n+1}X_n$ is at all times covered with functional oxygen (QO)-, hydroxyl (–OH)- or/or fluorine (–F)- groups, T_x [13]. For this reason, the chemical formula of MXenes becomes $M_{n+1}X_nT_x$. The ratios of different functional groups on MXene surfaces are in exact because they show variations with the differing etching conditions. The end-product, such as Ti_3C_2, combines exceptional physical and chemical properties and is exploited as catalyst, microwave absorber and electromagnetic interference (EMI) shield. Although it is similar to graphene, there are no data available on its resistance to corrosion [4, 14].

High-resolution electron microscopy and selected area electron diffraction studies have revealed that MXene are made up of orderly stacked layers with continuous crystal lattice edges [8]. Furthermore, the high-angle circular dark field spherical aberration-corrected scanning transmission electron microscope images have indicated that the transition metal and carbon layers are alternately arranged [8]. The size of these nanolayers exhibited resides between 0.5 and 200 nm with several nanometers in thickness [15].

MXene crystal structure generally exhibits a structure similar to that of MAX phase ceramics (Figure 4.1) [16, 17]. The MXenes of the ordered phase are more stable than the corresponding solid solution [7]. Transition metals are often arranged randomly in M layers, which are sandwiched between the layers of a second transition metal. The MXenes of the ordered phase are more stable than the corresponding solid solution [7, 17].

The very strong chemical bond between M and X enables MXene layers to display outstanding flexibility and durability performances [18] and better mechanical properties than those of 2D materials. These excellent mechanical features are provided by the delocalization of the d orbital electrons of M atoms and the surface stacking geometry together [19]. These electrons appear to be concentrated in the MX layer [16]. Multiple molecular dynamics (MD) calculations have demonstrated that MXene is several fold resistant to bending than graphene, and its bending stiffness can be several times higher than graphene [10, 18]. Here, it has been thought that the end groups

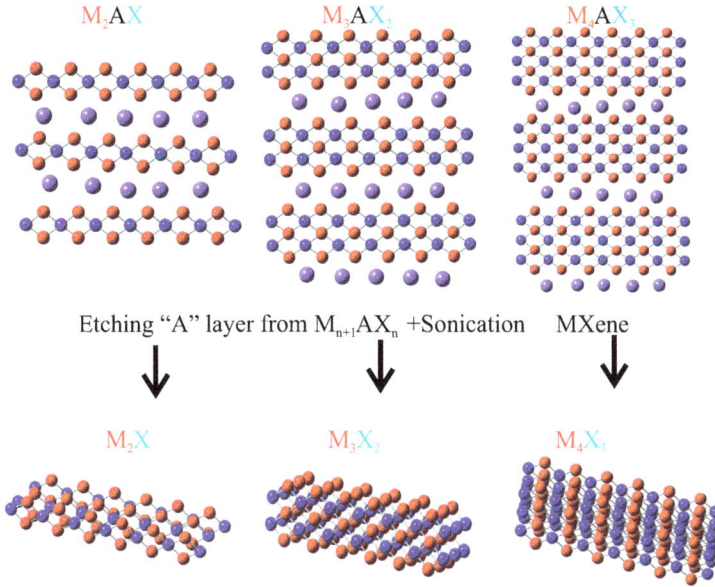

Figure 4.1: Sketched structure of MAX and MXene.

of MXene are responsible for the low elastic modulus but the unterminated MXene is rendered more flexible than graphene [12, 17].

4.2 MXene and MXene-based composites

Two-dimensional (2D) materials have appealed investigators since the discovery of graphene [2, 21] because 2D materials appeared to have suited to the electrochemical catalysis [20]. Lately, synthetic techniques have enabled the production of much larger 2D substances [5] and MXenes has been one of them [2, 6–22].

The resourceful MXene chemistry owes its existence to MXene's electronic, electrochemical, magnetic and mechanical characteristics. Its outstanding flexibility, stratified structure and 2D morphology enable it to serve as an excellent partner in composite materials. Therefore, MXenes and MXene-based composites have found the use in sodium-ion- and supercapacitor batteries [12, 24]. They have been especially exploited as efficient catalysts or cocatalysts [25] and photocatalytic reduction of carbon dioxide (CO_2) [14]; (ii) in the removal of heavy metal ions, organic dyes, eutrophic substances and nuclear waste from aqueous milieu [26]; (iii) in the production of sensors [27]. Several reviews have well-documented MXenes [28] and their applications [29] precluding MXene-based composites. As pollution has become a global concern, MXenes and MXene-based composites have gained further importance [23].

4.2.1 Production of MXene composites

4.2.1.1 Hydrothermal/solvothermal synthesis

The hydrothermal/solvothermal process involves a mineralizer, a liquid solvent and a precursor at high temperature and high pressure using water or organic solvents, respectively. It is a cost-effective means of secondary material synthesis [9]. An autoclave provides the reaction vessel in which the high pressure and high temperature conditions are achieved. Above boiling point a supercritical liquid phase, made up of a liquid and gas mixture, is obtained in the vessel and surface tension is thus eliminated at the interface [18]. This approach suits combining MXenes with transition metal oxides, nitrides, phosphides, perovskites, or chalcogenides [30]. One of its major drawbacks is the corrosive nature of the reaction process, which harm the desired chemical structure of the end-product [31].

4.2.1.2 Deposition methods

In generating MXene composites, deposition can be achieved by several means per se or in combination: electro deposition, chemical vapor deposition (CVD), atomic layer deposition (ALD), or photo deposition. In CVD, the substrate is decomposed with multiple volatile precursors in order to produce a thin film on its surface. The deposited material can be an alloy, metal, or a nanocomposite. CVD process can combine ALD to obtain two half reactions in order to prevent precursors from self-interaction. Here the progression of the ALD film is restricted and controllable. This allows a monolayer precipitation down to 0.1 Å in thickness. ALD procedure also enables the film to be covalently bonded on the substrate's surface. A successful ALD application, Pt–TBA–$Ti_3C_2T_x$, was shown below (Figure 4.2) [19, 31]. MXene hybrids are synthesized by electrodeposition using C-based materials, transition metal phosphites, oxides and metals. Photodeposition can be handy in the deposition of Cu or Pt on MXene surfaces. Photodeposition reactions can be controlled but it is relatively expensive [10, 12, 31].

4.2.1.3 Solution processing

Solution processing is one of the commonest means used for the synthesis of MXene-reinforced polymer composites because the formation of hydrophilic MXene nanosheets requires numbers of functional groups. Reaction media to be used should be polar such as dimethylsulfoxide, *N,N*-dimethylformamide, or bipolar like water [24]. This process can combine MXenes with a variety of polymeric substances such as polyurethane (PU), cellulose olyethylene oxide, polyvinyl alcohol, chitosan, polyacrylate acrylic resin polybenzimidazole, polyfluorenes, ethyleneimine, polyacrylamide

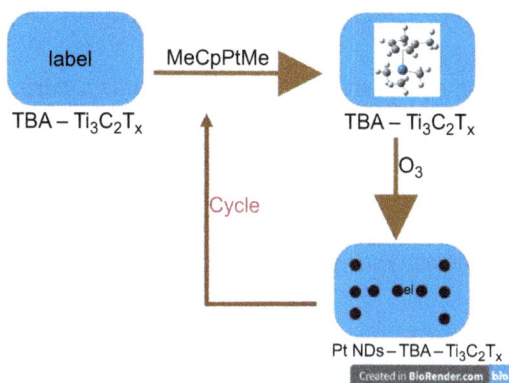

Figure 4.2: Schematic diagram of the synthesis of Pt–TBA–Ti$_3$C$_2$T$_x$ using ALD.

and polyvinylidene fluoride (PVDF) and with inorganic materials like TM oxides, chalcogenifriendly with dyes, phosphites and metal-organic frameworks (MOFs) [14, 32]. The end-products suffer some defects such as having poor mechanical features and not being very environmental-friendly as it involves inefficient evaporation steps [31].

4.2.1.4 Drop-casting and adsorption

In this approach the substances in the reaction mixture do not form covalent or ionic bonds; instead, they are held together by electrostatic and van der Waals interactions [33]. Therefore, the procedure does not involve high pressure and high temperature, and stable materials are interacted in mild conditions. It is very handy for the production of polymer composites [16, 31].

4.2.1.5 Hot press technique

MXenes display better thermal stability and require much higher decomposition temperatures than polymeric substances. In this solvent-free and temperature-controlled process, MXenes are often combined with different types of hydrophobic polymers such as LLDPE, PU, UHMWPE [34], polystyrene, PVDF, and polyaniline (PANI) [26] at a preferred density using a mixer [30]. High pressure is then applied to the molten mixture to obtain the final product [35]. It is thus cost-effective as it involves relatively fewer steps [31].

4.2.1.6 In-situ polymer blending

In this method, MXene nanoflakes are homogenously blended with monomers, curing agents and initiators using wet processes. Homogenous dispersion improves the quality of MXene nanosheets and provides them with outstanding interfacial strengths [35]. The end-products thus possess desired electrical, mechanical and thermal properties [36]. This process has been employed in the incorporation of polypyrrole [27], PANI [37] and polythiophene [38] into MXenes. A schematic representation of the process was provided below (Figure 4.3) [31].

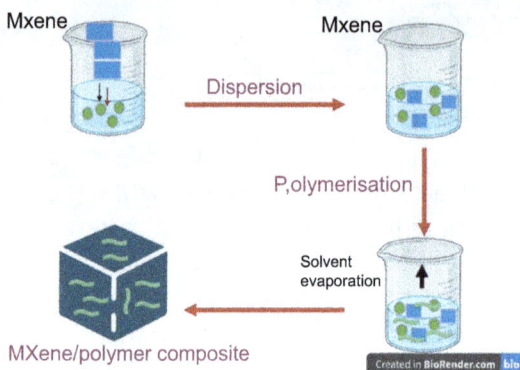

Figure 4.3: Schematic representation of the in situ polymerization technique.

4.2.2 MXene/polymer composite

Different MXene composites can be obtained using different matrix material: polymer matrix composites, metal/ceramic matrix composites, carbon composites and MXene hydrogels [39, 40]. The presence of surface functional groups such as –F, –O and –OH enables MXenes to bind polymeric materials better than graphene [35]. As the composite growth and crystallization behaviors are governed by MXene, the end-products are also often superior in terms of thermal stability, mechanical properties and electrical conductivity. $Ti_3C_2T_x$/polyethylene (PE), $Ti_3C_2T_x$/polyethylene glycol, $Ti_3C_2T_x$/PVDF, $Ti_3C_2T_x$/polyacrylonitrile, $Ti_3C_2T_x$/polybenzimidazole and $Ti_3C_2T_x$/polyvinyl alcohol composites have effectively been produced using this approach [31].

4.2.3 MXene-metal-ceramic composites

Metals besides being conductors also serve as chemical catalysts, and when incorporated, they improve such features of other materials. And the addition of MXene to ceramics or metals improves the overall mechanical properties of the end-product. MXenes' rich hydrophilic surfaces and excellent mechanical properties make them supreme underpinning vectors for metallic materials such as Cu and Al. In return, metal oxides/sulfides inhibit re-stacking of MXene layers and produce a synergistic effect. For this reason, various metal sulfides and metal oxides have been incorporated into MXenes [41]. As low as 2% MXene, $Ti_3C_2T_x$ could increase the mechanical strength of Al_2O_3 up to threefold [31].

4.2.4 MXene-carbon composites

Carbon-based materials, like carbon nanotubes (CNTs), carbon nanofibers (CNFs), graphene and porous carbon, could be incorporated into MXenes to enhance their electrochemical, conductivity and malleability performances [41]. In graphene/MXene composites, the added MXene can also prevent stacking of graphene layers almost without reducing the surface area and electrical properties [42].

CNFs, when combined with other materials, form a web through which electrons flow. These fibers can readily build bridge-like structures between the MXene layers. In the presence of Co catalyst, CNF can be loaded on the MXenes layers. The ends of the CNF become attached to neighboring MXene layers and form open frame-like structures. CNFs can thus increase the electrochemical performance of the end material. The CVD method has also been employed in the preparation of $Ti_3C_2T_x$/CNTs, during which CNTs are uniformly dispersed between MXene layers. The resulting compound exhibited excellent electromagnetic wave absorption (99.999% absorption), lower filler charge (35% wt) and wider absorption bandwidth (4.46 GHz), which can be increased up to 14.54 GHz by changing the thickness [31].

4.2.5 MXene-based hydrogels

Hydrogels is made up of water (99%) and inorganic particles, polymers and dissolved molecules. Incorporation of MXenes as nanofillers or as crosslinkers can increase tensile strength of hydrogels up to 10-fold and improve the conductivity of the end-product, the M-hydrogel [43]. Despite all the striking potentials, MXenes provide hydrogels with metastable characters. A schematic representation for a PVA-hydrogel generation process was shown below (Figure 4.4) [31].

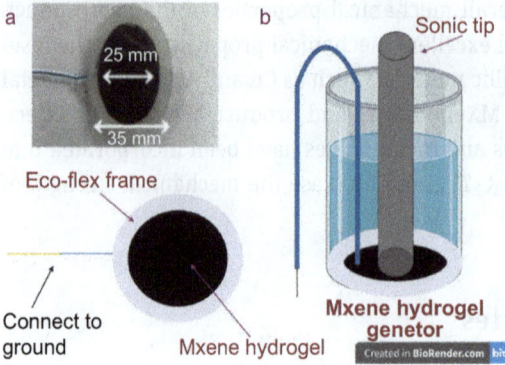

Figure 4.4: A simple PVA-hydrogel setup.

4.3 Industrial MXene nanocomposites

MXenes have received a significant medical interest. Since the appearance of the first MXene ($Ti_3C_2T_x$) in 2011, more than 30 MXene compounds have been produced including $Ti_3C_2T_x$, Ti_2CT_x, V_2CT_x, Nb_2CT_x, Ti_3CNT_x, $(Ti_{0.5}, Nb_{0.5})_2CT_x$, $Nb_4C_3T_x$ and $Ta_4C_3T_x$. A schematic representation of the MAX phase and MXene composition was provided below (Figure 4.5) [31].

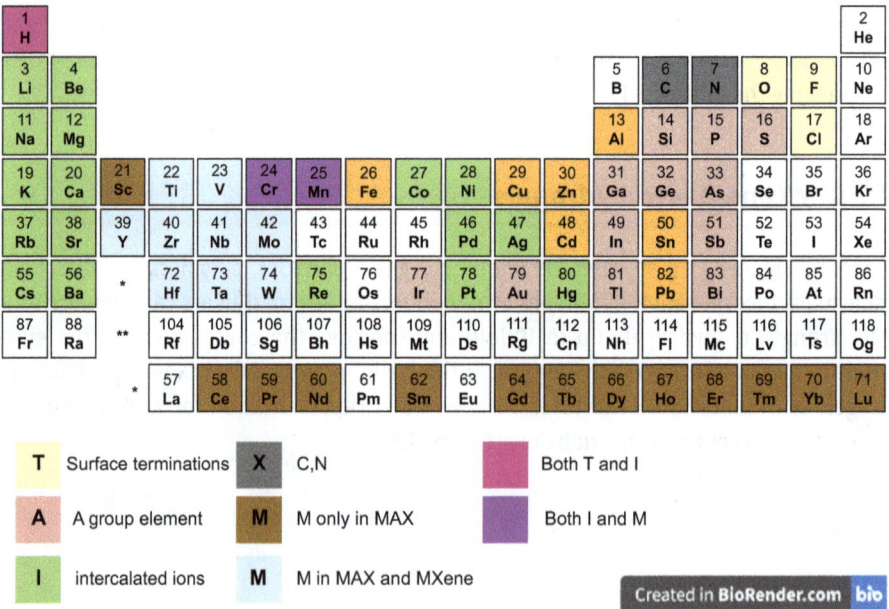

Figure 4.5: Periodic table depicting the MAX phase and MXene compositions.

Based on their structure, MXenes can be classified as mono-transition metal MXenes, dual-transition metal MXenes and hollow MXenes. Mono transition metal MXenes adopts three different frameworks in the M region with a transition metal such as M_2C, M_3C_2 and M_4C_3. Dual transition metal MXenes, consisting of two different transition metals giving rise to a new family, have the general formulas $M_4M'C_4$, $M_2M'_2C_3$ or $M_2M'C_2$, where M and M' represent two different transition metals. A few such as Cr_2TiC_2, $Mo_2Ti_2C_3$ and Mo_2TiC_2, Mo or Cr atoms occupy the outer boundaries of MXene and these atoms regulate its electrochemical properties. For others, such as $(Mo, V)_4C_3$ and Mo_4VC_4, the metals are randomly arranged for the whole structure in solid solutions. Also, in-plane chemical sequencing and selective etching have been used to synthesize MXene with well-ordered metal space [44].

More than 70 known MAX phases are available. The wide variety of MXene compositions and an immeasurable number of solid solutions provide us with a new set of properties, which can be fine-tuned by changing the proportions of the *M* or *X* elements (Figure 4.6) [31].

MXenes bring together the superior mechanical properties and electrical conductivity, fashioned by carbides and nitrides of transition metals in the presence of surface oxygen or hydroxyl moieties. Their large negative zeta potential, ceramic nature and ability to form a stable colloid in water make them suitable for a wide variety of applications [44]. The tunability of their physicochemical properties opens possibilities for

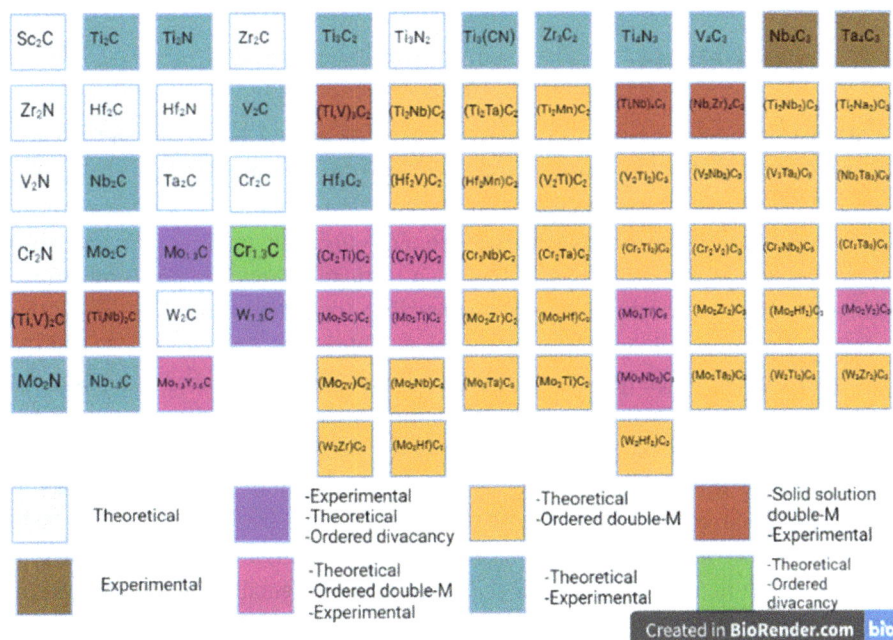

Figure 4.6: Experimental (blue) and theoretical (gray) MXenes.

developing functional compounds in combination with other nanomaterials. Computational studies have also revealed that MXenes can strongly interact with noble metals such as Pd, Pt, Au and Ru [31].

Experimental MXenes have been used as supercapacitors, Li-ion batteries, catalysts, transparent conductors, biomaterials, lubricants, field-effect transistors, sensors, drug carriers, dual-sensitivity surfaces, EMI shielding materials, purifiers and polymeric fillers. They have found applications in composites, hybrid nanocomposites, dye substrates and cancer therapy. Moreover, MoS_2-functionalized MXene finds application in mercury absorption.

Like most two-dimensional compounds, MXenes undergo restacking, which restrict its application. Many inorganic and organic compounds such as polymers, carbon-based materials, transition metal-based compounds, quantum dots (QDs), metal alloys and MOFs have successfully been integrated with MXenes to produce hybrid compounds. These MXene-based hybrids have proven to be more stable than their individual components and have enabled the diversification of the application areas [30]. Integration of 2D MXenes into 3D structures could also greatly inhibit re-stacking phenomenon. 3D MXenes could be useful in sensor applications. However, one major challenge in applying MXenes to industrial applications is the stability, as the presence of a large number of surface groups makes them highly susceptible to oxidative degradation [31].

4.4 Conclusions

2D MXenes have been proven to be outstanding anticorrosive materials, and this has opened a new era for high-performance anticorrosion coatings. MXene composites have extensively been used as epoxy and PU fillers in polymers. Studies have proven that the presence of MXenes and their compounds effectively improves the anticorrosion homes in addition to the sturdiness of such coatings. Although those 2D substances had been broadly researched over the past few decades, MXenes and their composites nonetheless face several demanding situations of their realistic application. One of the not unusual place-demanding situations of the use of MXenes and their composites as nanofillers in polymer coatings is their nonuniform dispersion in polymer matrices because of their aggregation properties. Fortunately, this project may be triumph over with the aid of using ultrasonic coating formulations. Another issue in the use of MXenes and their compounds is that they had been synthesized the use of very high-priced substances and aggressive chemicals. Therefore, MXens and their composite syntheses are each high-priced and now no longer environmentally friendly. Therefore, there's immoderate call to expand cost-powerful and environmentally pleasant strategies for the synthesis of MXenes and their compounds. MXenes are inherently noticeably hydrophilic because of the presence of polar floor useful groups; however, their aqueous segment programs as corrosion inhibitors are restricted because of their capacity to oxidize and degrade in water. Therefore, future studies should focus on the design and synthesis of water-resistant MXenes to be used as aqueous phase anticorrosive materials. Expanding the MXenes family to identify the new $M_{n+1}X_nT_x$ with various functions is essential for the development of MXenes nanocomposite materials. Among the MXenes, $Ti_3C_2T_x$ is most commonly used, so the use of other MXenes such as V_2CT_x, Mo_2C, V_2C, Nb_2C and V_4C_3 should be explored.

References

[1]   Zhang, B., et al., Facile fluorine-free one step fabrication of superhydrophobic aluminum surface towards self-cleaning and marine anticorrosion, Chemical Engineering Journal, 2018, 352, 625–633.

[2]   Zhang, B., et al., Controllable Dianthus caryophyllus-like superhydrophilic/superhydrophobic hierarchical structure based on self-congregated nanowires for corrosion inhibition and biofouling mitigation, Chemical Engineering Journal, 2017, 312, 317–327.

[3]   Eduok, U., et al., Fabricating protective epoxy-silica/CeO2 films for steel: Correlating physical barrier properties with material content, Materials & Design, 2017, 124, 58–68.

[4]   Yan, H., et al., Ti3C2 MXene nanosheets toward high-performance corrosion inhibitor for epoxy coating, Progress in Organic Coatings, 2019, 135, 156–167.

[5]   Liu, J., et al., Silane modification of titanium dioxide-decorated graphene oxide nanocomposite for enhancing anticorrosion performance of epoxy coatings on AA-2024, Journal of Alloys and Compounds, 2018, 744, 728–739.

[6]   He, Y., et al., Improved corrosion protection of waterborne epoxy/graphene coating by combining non-covalent and covalent bonds, Reactive and Functional Polymers, 2019, 137, 104–115.

[7]   Chen, C., et al., Achieving high performance corrosion and wear resistant epoxy coatings via incorporation of noncovalent functionalized graphene, Carbon, 2017, 114, 356–366.

[8]   Pourhashem, S., et al., Excellent corrosion protection performance of epoxy composite coatings filled with amino-silane functionalized graphene oxide, Surface and Coatings Technology, 2017, 317, 1–9.

[9]   Li, X., et al., Enhanced lithium and electron diffusion of LiFePO4 cathode with two-dimensional Ti3C2 MXene nanosheets, Journal of Materials Science, 2018, 53(15), 11078–11090.

[10] Rasool, K., et al., Antibacterial activity of Ti3C2T x MXene, ACS Nano, 2016, 10(3), 3674–3684.

[11] Zhang, C., et al., h-BN decorated with Fe3O4 nanoparticles through mussel-inspired chemistry of dopamine for reinforcing anticorrosion performance of epoxy coatings, Journal of Alloys and Compounds, 2016, 685, 743–751.

[12] Come, J., et al., Nanoscale elastic changes in 2D Ti3C2Tx (MXene) pseudocapacitive electrodes, Advanced Energy Materials, 2016, 6(9), 1502290.

[13] Hantanasırısakul, K., et al., Fabrication of Ti3C2Tx MXene transparent thin films with tunable optoelectronic properties, Advanced Electronic Materials, 2016, 2(6), 1600050.

[14] Xu, G., et al., Solvent-regulated preparation of well-intercalated Ti3C2Tx MXene nanosheets and application for highly effective electromagnetic wave absorption, Nanotechnology, 2018, 29(35), 355201.

[15] George, S.M., Kandasubramanian, B., Advancements in MXene-Polymer composites for various biomedical applications, Ceramics International, 2020, 46(7), 8522–8535.

[16] Michael, N., et al., Two-dimensional materials: 25th anniversary article: MXenes: A new family of two-dimensional materials (Adv. Mater. 7/2014), Advanced Material, 2014, 26(7), 992–1005.

[17] Chen, X., et al., MXene/polymer nanocomposites: Preparation, properties, and applications, Polymer Reviews, 2021, 61(1), 80–115.

[18] Kurtoglu, M., et al., First principles study of two-dimensional early transition metal carbides, Mrs Communications, 2012, 2(4), 133–137.

[19] Fu, Z., et al., Mechanistic quantification of thermodynamic stability and mechanical strength for two-dimensional transition-metal carbides, The Journal of Physical Chemistry C, 2018, 122(8), 4710–4722.

[20] Lı, L., et al., Black phosphorus field-effect transistors, Nature Nanotechnology, 2014, 9(5), 372–377.

[21] Wang, H., Maiyalagan, T., Wang, X., Review on recent progress in nitrogen-doped graphene: Synthesis, characterization, and its potential applications, ACS Catalysis, 2012, 2, 781–794.

[22] Khazaeı, M., et al., Electronic properties and applications of MXenes: A theoretical review, Journal of Materials Chemistry C, 2017, 5(10), 2488–2503.

[23] Zhan, X., et al., MXene and MXene-based composites: Synthesis, properties and environment-related applications, Nanoscale Horizons, 2020, 5(2), 235–258.

[24] Yan, J., et al., Flexible MXene/graphene films for ultrafast supercapacitors with outstanding volumetric capacitance, Advanced Functional Materials, 2017, 27(30), 1701264.

[25] Guo, Z., et al., MXene: A promising photocatalyst for water splitting, Journal of Materials Chemistry A, 2016, 4(29), 11446–11452.

[26] Zhang, Q., et al., Efficient phosphate sequestration for water purification by unique sandwich-like MXene/magnetic iron oxide nanocomposites, Nanoscale, 2016, 8(13), 7085–7093.

[27] Yu, X., et al., Monolayer Ti_2CO_2: A promising candidate for NH_3 sensor or capturer with high sensitivity and selectivity, ACS Applied Materials & Interfaces, 2015, 7(24), 13707–13713.

[28] Ng, V.M.H., et al., Recent progress in layered transition metal carbides and/or nitrides (MXenes) and their composites: Synthesis and applications, Journal of Materials Chemistry A, 2017, 5(7), 3039–3068.

[29] Nan, J., et al., Small, 2019, e1902085.

[30] Lim, K.R.G., et al., Rational design of two-dimensional transition metal carbide/nitride (MXene) hybrids and nanocomposites for catalytic energy storage and conversion, ACS Nano, 2020, 14(9), 10834–10864.

[31] Prakash, N.J., Kandasubramnian, B., Nanocomposites of MXene for industrial applications, Journal of Alloys and Compounds, 2021, 862, 158547.

[32] Tian, W., et al., Multifunctional nanocomposites with high strength and capacitance using 2D MXene and 1D nanocellulose, Advanced Materials, 2019, 31(41), 1902977.

[33] Ng, V.M.H., et al., Correction: Recent progress in layered transition metal carbides and/or nitrides (MXenes) and their composites: Synthesis and applications, Journal of Materials Chemistry A, 2017, 5 (18), 8769–8769.

[34] Zhang, H., et al., Preparation, mechanical and anti-friction performance of MXene/polymer composites, Materials & Design, 2016, 92, 682–689.

[35] Fang, R.H., et al., Cell membrane coating nanotechnology, Advanced Materials, 2018, 30(23), 1706759.

[36] Carey, M., et al., Nylon-6/Ti_3C_2T z MXene nanocomposites synthesized by in situ ring opening polymerization of ε-caprolactam and their water transport properties, ACS Applied Materials & Interfaces, 2019, 11(22), 20425–20436.

[37] Ren, Y., et al., Synthesis of polyaniline nanoparticles deposited on two-dimensional titanium carbide for high-performance supercapacitors, Materials Letters, 2018, 214, 84–87.

[38] Liu, R., et al., Ultrathin biomimetic polymeric Ti_3C_2T x MXene composite films for electromagnetic interference shielding, ACS Applied Materials & Interfaces, 2018, 10(51), 44787–44795.

[39] Patıl, N.A., Njuguna, J., Kandasubramanian, B., UHMWPE for biomedical applications: Performance and functionalization, European Polymer Journal, 2020, 125, 109529.

[40] Rezaeı, B., Lotfiforushani, H., Ensafi, A.A., Modified Au nanoparticles-imprinted sol–gel, multiwall carbon nanotubes pencil graphite electrode used as a sensor for ranitidine determination, Materials Science and Engineering: C, 2014, 37, 113–119.

[41] Yang, J., et al., MXene-based composites: Synthesis and applications in rechargeable batteries and supercapacitors, Advanced Materials Interfaces, 2019, 6(8), 1802004.

[42] Wang, B., et al., In situ synthesis of Co3O4/graphene nanocomposite material for lithium-ion batteries and supercapacitors with high capacity and supercapacitance, Journal of Alloys and Compounds, 2011, 509(29), 7778–7783.

[43] Zhang, Y.Z., et al., MXenes stretch hydrogel sensor performance to new limits, Science Advances, 2018, 4(6), eaat0098.

[44] Gogotsi, Y., Anasori, B., The rise of MXenes, ACS Nano, 2019, 13(8), 8491–8494.

Bo-kai Liao*, Zhi-Gang Luo, Shan Wan, Hao-Wei Deng, Shu-Yi Jiang,
Shuang-Jian Li, Jun-Jie Yang

Chapter 5
Quantum dots in corrosion prevention

Abstract: Metal corrosion is a natural spontaneous behavior in the earth environment, which greatly reduces the service lifetime in the industrial occasions and even causes potential safety hazard. The application of corrosion inhibitor is regarded as one of the most convenient and efficient methods for corrosion control. Quantum dots, as the typical zero-dimensional nanomaterial, have the unique chemical structure and possess broad application prospects in the field of anticorrosion. In this paper, the synthesis method, performance and mechanism of quantum dots as corrosion inhibitor are summarized in detail. The advantages and limitations for quantum dots as corrosion inhibitor are introduced, and the future development for this new type corrosion inhibitor has been forecasted.

Keywords: Corrosion, quantum dots, corrosion inhibitor

5.1 Background

Metal corrosion is a spontaneous natural phenomenon, which causes the huge damage to the service lifetime of device and buildings [1–3]. Corrosion type can be divided into uniform and localized corrosion, consisting of pitting corrosion, crevice corrosion, gal-

*Corresponding author: Bo-kai Liao, School of Chemistry and Chemical Engineering, Guangzhou University, Guangzhou 510006, China; Joint Institute of Guangzhou University & Institute of Corrosion Science and Technology, Guangzhou University, Guangzhou 510006, China,
e-mail: bokailiao@gzhu.edu.cn
Zhi-Gang Luo, School of Chemistry and Chemical Engineering, Guangzhou University, Guangzhou 510006, China
Shan Wan, School of Chemistry and Chemical Engineering, Guangzhou University, Guangzhou 510006, China; Joint Institute of Guangzhou University and Institute of Corrosion Science and Technology, Guangzhou University, Guangzhou 510006, China
Hao-Wei Deng, School of Physics and Materials Science, Center of Advanced Functional Materials, Guangzhou University, Guangzhou 510006, China
Shu-Yi Jiang, School of Art and Design, Guangdong University of Technology, Guangzhou 510062, China
Shuang-Jian Li, Institute of New Materials, Guangdong Academy of Sciences, National Engineering Laboratory of Modern Materials Surface Engineering Technology, Guangzhou 510650, China
Jun-Jie Yang, Institute of Advanced Wear and Corrosion Resistant and Functional Materials, Jinan University, Guangzhou 510632, China

https://doi.org/10.1515/9783111071756-005

vanic corrosion, intergranular corrosion, stress corrosion cracking, etc. [2, 4–7]. To make corrosion rate-controllable within an acceptable range, numbers of methods have been built like cathodic protection, sacrificial anode, coating [8–11] and corrosion inhibitor [12–14]. Among them, the usage of corrosion inhibition can be regarded as the most convenient corrosion protection method, and large numbers of corrosion inhibitor have been developed in past decades [15, 16]. Traditional corrosion inhibitor can be classified as organic and inorganic types [17–20], and inorganic-type corrosion inhibitors mainly consist of inorganic salts like molybdate, nitrite and silicate. Organic-type corrosion inhibitors are composed of heterocyclic substances like imidazole derivatives, azazole derivatives and pyrimidine derivatives. [21–24]. Both types of corrosion inhibitor can achieve satisfactory corrosion protection performance with a suitable dosage. However, some toxic and harmful components bring huge harm to the environmental and human health [25, 26]. Many countries have banned the use of some toxic corrosion inhibitor [27, 28], for example, hexavalent chromium possesses excellent anticorrosion capability; however, it can also induce some irreversible health damage like skin irritation and lung cancer [29].

In this case, numerous eco-friendly corrosion inhibitors have been prepared including natural product-extracted corrosion inhibitor [30–33] and ionic liquid [34–37]. Despite these eco-friendly corrosion inhibitors overcome some shortcomings, they cannot completely replace the traditional corrosion inhibitors [38, 39] like unknown active compositions [40–42], poor long-term corrosion inhibition performance [43, 44] and reproducibility [45]. To date, many researchers are still seeking or developing some other new green corrosion inhibitors. Recently, materials in nanoscale attract wild attentions due to their larger specific surface area [46], rich chemical groups and the decreasing cost like nanodot [47], nanosheet [48–50] and metal-organic framework [51–53]. In recent years, some carbon-based nanomaterials have been successfully synthesized and utilized as corrosion inhibitors in solution media and coating systems [54] like carbon dots [54–56], carbon nanotube [57] and graphene [58–60]. Among them, carbon quantum dots (CDs) is first discovered since 2004, and the functionalized carbon quantum dots (FCDs) as corrosion inhibitors display the superior anticorrosion performance, which can act as the effective corrosion inhibitors in acidic solutions [61, 62]. Huang and coworkers [63] demonstrated a new micro area electrochemical stripping method for large-scale preparation of nitrogen-doped CDs, whose corrosion inhibition efficiency of copper was 96.32% at 150 mg/L. Li and coworkers [64] used citric acid as carbon source and diethylenetriamine and urotropine as precursors to synthesize N-doped CDs with luminescent properties by microwave method. The results showed that this functionalized CDs had good corrosion inhibition performance at 60 °C, which can reach 81.2% when the dosage was 600 mg/L. Wei and coworkers [65] adopted a robust purification process to synthesize a new nitrogen-doped CDs with an average size of 2.5 ± 0.8 nm through the hydrothermal reaction of aminoguanidine and citric acid, and the maximum inhibition rate was 95.3% when the concentration was 200 mg/L. Zhu et al. [66] synthesized N-doped CDs by hydrothermal method with folic acid and o-phenylenediamine as precursors, the inhibition efficiency of 150 mg/L N-CDs reached 95.4%. N-doped CDs adsorbed on the steel surface by coordinating its electron-rich atoms with iron metal to form a protective film, thus slowing down the dissolution reaction of anode metal and achieving corrosion inhibition. Kalajahi et al. [67] reported that copper nanoparticle-doped CDs (Cu-CDs) nanocomposites can be used as high-efficient corrosion inhibitors to reduce the microbiologically influenced corrosion. Guo and coworkers [68] first synthesized N and S codoped carbon dots (N, S-CDs) and used them as corrosion inhibitors. The low concentration is 10 mg/L N, S-CDs, and the corrosion current density is reduced. The inhibition efficiency is 93% when the concentration is 50 mg/L N and the concentration is N, S-CDs. Guo and coworkers [69] prepared N, S-CDs as corrosion inhibitors for aluminum alloys. The results confirmed that N, S-CDs can adhere to the aluminum surface through the diffusion and condensation effects of nanomaterials, and the interaction between aluminum ions and N, S-CDs led to the formation of a protective network corrosion inhibitor film on the aluminum surface. Zhang et al. [70] studied the synthesis of new N-doped carbon dots from citric acid and L-serine, and the obtained N-CDs was an effective inhibitor with inhibition efficiency about 98.5% after soaking for 24 h, which interacted with copper matrix through chemical and physical adsorption. Zhang et al. [71] also prepared a new type of N, S-CDs, with N and S content as high as 17% and 19%, respectively. In 0.5 mol/L H_2SO_4 solution, at a concentration of 50 mg/L, the new nanomaterials showed excellent effects on copper (99.88%). Yadav and coworkers [72] synthesized a new environment-friendly, water-soluble, and efficient corrosion inhibitor N, S-CDs by hydrothermal method. N, S-CDs showed 98.64% inhibition efficiency at 200 mg/L concentration and 303 K. Yadav and coworkers [73] synthesized CDs with environmental protection, good water solubility and low price, which showed 97.89% at a concentration of 100 mg/L in 15% HCl solution and a temperature of 303 K. Ye and coworkers [74] prepared three N-doped carbon point (N-CDs) inhibitors through natural lemon. And results showed that the inhibition efficiencies for these N-CDs at 200 mg/L were higher than 90%. Zhao and coworkers [75] prepared a series of N-CDs from tryptophan, and the optimum performance was obtained at 180 °C and 1 h. At 200 mg/L N-CDs, the inhibition efficiency to steel is higher than 90%. Guo and coworkers [76] synthesized melamine-modified carbon dots by hydrothermal method. The results showed that the maximum corrosion inhibition efficiency was 92.36% at 200 mg/L. The adsorption of Me-CDs conforms to the Temkin adsorption model. Qiang and coworkers [77] prepared environment-friendly nitrogen-doped CDs and nitrogen sulfur co-doped CDs through one-step hydrothermal process, which reached about 87.9% and 96.4% at 200 mg/L. Therefore, the simple preparation, eco-friendly and high-corrosion inhibition performance of N, S-CDs will provide a new way to design efficient carbon spots and broaden the application of carbon spots in the corrosion field.

In this work, recent advances on the application of quantum dots in corrosion protection field have been summarized. The chapter introduces the preparation method, model of adsorption kinetic, and anticorrosion mechanism for quantum dots in detail and look forward to its future development trend.

5.2 Preparation method

Many kinds of synthesis methods of quantum dots have been reported in recent decades, and it is an important hotspot to find environmentally friendly, large-scale and low-cost synthesis methods. Taking the preparation of CDs as an example, we prepared the glucose-based FCDs using one-step hydrothermal method (as shown in Figure 5.1), which displayed excellent anticorrosion property for Q235 carbon steel in acidic solution. Cui and coworkers [78] reported that CDs can be rapidly synthesized by simple hydrothermal method using lignin as raw material, the synthesized CDs can emit blue photoluminescence and have good light stability, good biocompatibility, low cytotoxicity and high water solubility. This work provides a new method for preparing CDs from natural materials and showed the potential of CDs in anticorrosion applications. Similarly, Wu et al. [79] utilized biomass lignin as carbon source, fluorescent carbon quantum dots were synthesized by a simple two-step method, and it also provides a theoretical basis for exploring the formation mechanism of CDs. The next step after selecting the appropriate synthesis method of CDs is to study the efficient utilization of CDs. Zhu et al. [80] synthesised CDs using the two-step route, which can be effectively used as fluorescent nanoprobe with high sensitivity and high selectivity for detecting iron ions. Li and coworkers [81] prepared CDs using alkali lignin (AL) doped with nitrogen. After hydrothermal treatment of AL and m-phenylenediamine, AL retains the original lignin skeleton and can prepare carbon quantum dots with good photoluminescence properties. Si and coworkers [82] used sulfated lignin as the substitute of phenol to prepare new lignin nanospheres by simple hydrothermal method. The prepared nanospheres were then used as reducing agents and stable carriers for synthesizing Pd nanoparticles, which provided a new way for the development of CDs.

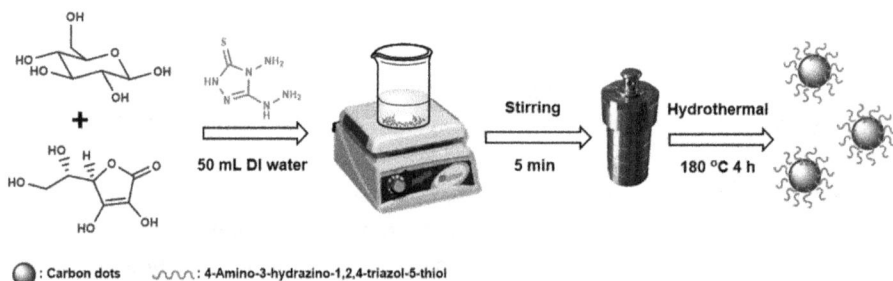

● : Carbon dots   ∿∿∿ : 4-Amino-3-hydrazino-1,2,4-triazol-5-thiol

Figure 5.1: The schematic diagram of preparation process for glucose-based functionalized carbon dots using hydrothermal method.

After preparation of CDs, some surface characterizations are carried out to measure the microstructure and surface chemical property such as transmission electron microscope (TEM), X-ray diffraction meter (XRD), Fourier transform infrared (FTIR) and

X-ray photoelectron spectroscopy (XPS). As shown in Figure 5.2, the size of CDs is usually about a few nanometers to tens of nanometers.

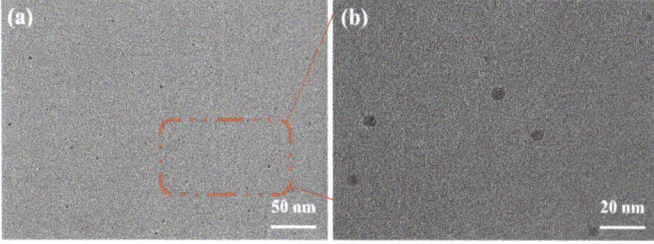

Figure 5.2: The TEM images for glucose-based functionalized carbon dots.

5.3 Evaluation method of corrosion inhibition behavior of quantum dot inhibitor

The corrosion inhibition of quantum dot inhibitor can be measured using weight loss and electrochemical techniques such as open-circuit potential, electrochemical impedance (EIS), potentiodynamic polarization curve (PDP) and scanning vibration electrode technique (SVET) as well as some surface characterization methods like scanning electron microscopy (SEM), energy-dispersive spectrometer, water contact angle, atomic force microscope (AFM), XDS and XPS. Generally, the corrosion inhibition efficiency can be calculated based on the weight loss method (as shown in eqs. (5.1) and (5.2)), EIS (as shown in eq. (5.3)) and PDP tests (as shown in eq. (5.4)):

$$v_{corr} = \frac{\Delta W}{S \times t} \tag{5.1}$$

$$\eta = \frac{v^0_{corr} - v_{corr}}{v^0_{corr}} \times 100\% \tag{5.2}$$

where ΔW means the mass loss (g) of the samples before and after weight loss test, S indicates the exposed area of metal (m^2), t is the time (h) for immersion test and v_{corr} and v^0_{corr} correspond the corrosion rates in the presence and absence of corrosion inhibitor, separately:

$$\eta_{PDP} = \frac{i^0_0 - i^0}{i^0_0} \tag{5.3}$$

where i^0_0 and i_0 are the corrosion current density calculated from PDP results in the absence and presence of corrosion inhibitor:

$$\eta_{\text{EIS}} = \frac{R_{\text{ct}} - R_{\text{ct}}^0}{R_{\text{ct}}} \times 100\% \qquad (5.4)$$

where R_{ct} and R_{ct}^0 are the charge transfer resistance obtained from EIS results in the presence and absence of corrosion inhibitor.

SVET method can be utilized as one in situ method to monitor the initiation and development of pit in the presence and absence of corrosion inhibitor. As shown in Figure 5.3, with the addition of CD corrosion inhibitor, the surface potential difference became smaller, indicating that this corrosion inhibitor can greatly retard the corrosion behavior. Besides, with the prolongation of immersion time, the surface potential difference in the blank solution became larger while it can remain at a low level, meaning this corrosion inhibitor showed the superior long-term anticorrosion performance. To gain the spatial and temporal corrosion electrochemical response signal,

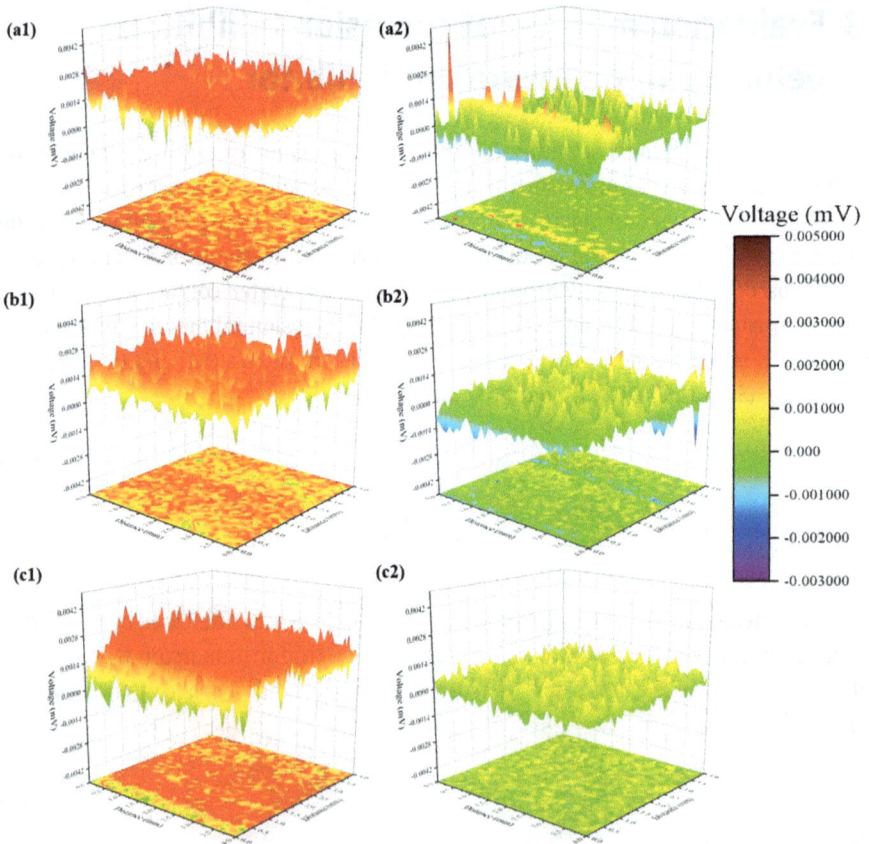

Figure 5.3: Potential difference distribution of SVET test for Cu in 3.5 wt% NaCl solution without and with glucose-based functionalized carbon dot inhibitor after different immersion time: 1 h (a1, a2), 6 h (b1, b2) and 24 h (c1, c2).

some other techniques need to be developed to real time in situ to observe the anti-corrosion process of corrosion inhibitor.

After corrosion test, the metal surface is always observed to illustrate the anticorrosion mechanism of CDs corrosion inhibitor including corrosion morphology, hydrophobicity, roughness and surface adhesion. As shown in Figure 5.4, in the presence of CDs corrosion inhibitor, the surface morphology became smoother and less corrosion products can be found from SEM images (Figure 5.4a and b), meaning that the corrosion behavior was slowed down by the inhibitor. Meanwhile, the surface adhesive force was measured using AFM force–distance method (Figure 5.4c), which increased compared with that in blank solution (Figure 5.4d) and proved the existence of adsorption film on metal surface. With the adsorption of inhibitor molecule, the surface hydrophobicity was significantly changed, and it generally became more hydrophobic and can thus retard the invasion of some water-soluble aggressive ions, which is beneficial for slowing down the corrosion behavior.

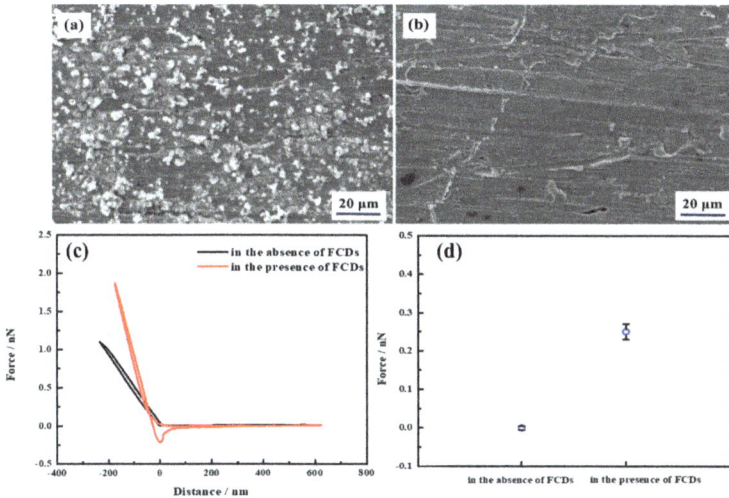

Figure 5.4: (a and b) SEM images for metal surface morphology and (c and d) force–distance curves and forces measured on the Cu samples after 24 h immersion in 3.5 wt% NaCl solution with and without glucose-based functionalized carbon dots inhibitor.

5.4 Adsorption behavior of quantum dot inhibitor

Many adsorption isotherm models have been proposed to research the adsorption behavior of corrosion inhibitor molecule [14] including Langmuir model [83–85], Temkin model [86–88], Frumkin model [89–91], Freundlich model [92–94], El-Awady model, Brockris Swinkel and Flory–Huggins adsorption isotherm models, as shown in eqs. (5.5)–(5.8):

Langmuir model:

$$\frac{C}{\theta} = \frac{1}{K_{ads}} + C \qquad (5.5)$$

Temkin model:

$$e^{-2a\theta} = KC \qquad (5.6)$$

Frumkin model:

$$\ln\left[\frac{\theta}{(1-\theta)C}\right] = \ln K + 2a\theta \qquad (5.7)$$

Flory–Huggins model:

$$\ln\frac{\theta}{C} = x\ln(1-\theta) + \ln(xK_{ads}) \qquad (5.8)$$

Based on these adsorption isotherm models, which combine the surface coverage of corrosion inhibitor molecule with the equilibrium constant of adsorption, the adsorption mechanism can be proposed [76, 95, 96]. Among them, the surface coverage is equal to the corrosion inhibition efficiency. When the adsorbate-containing phase is in contact with an adsorbent for a sufficient time, the adsorption equilibrium can be established, and the obtained regression coefficient can be used as a parameter to choose the proper model [97]. Take the melamine-modified carbon dots as an example, we compared the abovementioned adsorption isotherm models (eqs. (5.5)–(5.8)) and found that the Temkin adsorption isotherm model had the best correlation (R^2 = 0.99176). In this case, we used this model to describe the adsorption behavior of melamine-modified carbon dots for carbon steel in NaCl solution.

Under the premise of knowing the molar weight of corrosion inhibitor molecule, the adsorption free energy (ΔG_{ads}) can be calculated with the obtained K_{ads} from adsorption isotherm model according to the following equation:

$$\Delta G_{ads} = -RT\ln(1,000K_{ads}) \qquad (5.9)$$

where R is equal to 8.314 J/mol K, meaning the molar gas constant and T present the temperature in kelvin. Based on the calculated value of ΔG_{ads}, researchers can identify the adsorption type of corrosion inhibitor molecular including physical adsorption (more positive than −20 kJ/mol), chemical adsorption (more negative than −40 kJ/mol) and physiochemical adsorption (between −20 and 40 kJ/mol). The physical adsorption process contains the mutual Coulomb force between the unlike charges including corrosion inhibitor molecule and charged metal surface. And the chemical adsorption is associated with the formation of chemical bond due to the charge transfer or sharing

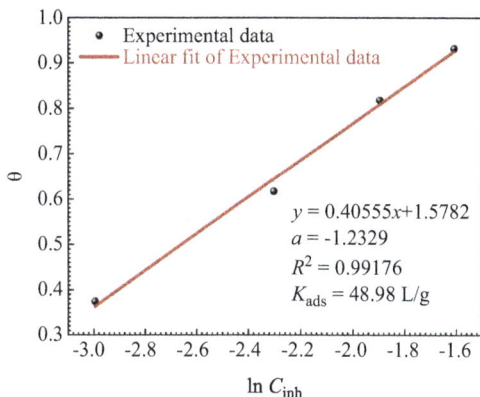

Figure 5.5: Temkin adsorption isotherm model of melamine-modified carbon dots for Q235 carbon steel immersed in 3.5 wt% NaCl solution [76].

between corrosion inhibitor molecule and metal atom. The physiochemical adsorption includes both the processes. The thermodynamic parameters including entropy $\left(\Delta s^0_{ads}\right)$ and enthalpy $\left(\Delta H^0_{ads}\right)$ can be further calculated using eq. (5.10) to depict the spontaneity and change of disorder during corrosion inhibitor adsorption process:

$$\ln K_{ads} = -\frac{\Delta H^0_{ads}}{RT} + \frac{\Delta s^0_{ads}}{R} + \ln\frac{1}{55.5} \tag{5.10}$$

Cui et al. [98, 99] found that the adsorption processes of *p*-phenylenediamine, *o*-phenylenediamine and N, S co-doped CDs obey Langmuir adsorption isotherm, which both involve physisorption and chemisorption according to the results of adsorption free energy. Ye et al. [100] attributed the excellent anticorrosion property of functionalized citric acid-based CDs to the coverage of adsorption film on metal surface, which obeys to the Langmuir model and belongs to the physicochemical interaction. Yang et al. [101] also found that the spontaneous adsorption process for the functionalized citric acid-based CDs using imidazole obeys to the Langmuir model, and the adsorption mechanism is a combination of physical and chemical adsorption. Saraswat and Yadav [72] indicated that N, S-CD inhibitor follows the Langmuir adsorption isotherm and can act as the mixed-type adsorption inhibitor. The work proposed some adsorption isothermal models for CD inhibitors, which are restricted for monolayer or multilayer adsorption at molecular scale. As for CD inhibitors at nanoscale, this limitation must be paid enough attention. Cen et al. [102] proposed the Redlich-Peterson (R-P) equation for the adsorption of functionalized carbon nanotube with two hypotheses as basis, which contained the properties of Langmuir and Freundlich models. Moreover, it is noteworthy that the accurate molar weight of CD inhibitor molecule should be measured to use eq. (5.9), and the synthesized CDs are usually with different molar weights, which can be measured by inductively coupled plasma mass spectrometry.

5.5 Fundament of corrosion inhibition mechanism

Traditional inorganic and organic-type corrosion inhibitors, as the dispersed phase, can be quickly and uniformly dispersed in the dispersion media including water or oil. In this case, these corrosion inhibitor molecules with less molecular weight can easily transfer from solution to metal surface and then reach a stable adsorption state on the metal surface, forming a complete protective film in a short time. Due to the relatively larger mass of quantum dots not only the mass transfer process is relatively slower from solution bulk to metal surface but also the adsorption behavior for quantum dots is slower. Therefore, it needs to take longer time to adsorb on metal surfaces. As reported in our previous work [10, 76], the desorption and adsorption process reached the stable state at about 6 h, which is largely longer compared with that of the organic or inorganic inhibitor molecule. The formation of protective film is affected by the amount of CDs, which is limited by the mass transfer process in solution as well as by the structure and solution chemistry [103]. The processes have not been reported about the mass transfer and film-forming. We tried to utilize the real-time quartz crystal microbalance measurement combined with electrochemical tests to investigate the adsorption behavior of nanomaterial corrosion inhibitor on metal surface, including CDs, but the inductive signal from weight loss during corrosion dramatically affect the weight change during the adsorption of nanomaterial corrosion inhibitor. Some other in situ methods can be built to clarify the difference of nanomaterial corrosion inhibitor with traditional corrosion inhibitor molecules.

Moreover, due to the large specific surface area and the easy accumulation of charge for nanoparticles, the nanoparticles in the solution tend to generate strong interaction forces, making the particles easy to aggregate [104]. The agglomeration effect of nanoparticles makes it difficult to disperse uniformly in the system, thus reducing the corrosion inhibition effect. However, the agglomeration behavior is beneficial to quantum dots, which can accumulate and deposit while adsorbing on the metal surface, eventually forming a thicker protective film than the traditional corrosion inhibitor adsorption film, which has a relatively stronger corrosion inhibition effect. The surface of the quantum dots can contain a large number of active groups, such as hydroxyl and carboxyl, and then the modified quantum spot doped with heteroatoms can form a coordination compound with the outer layer of the metal, thus having strong adsorption on the metal surface [105]. As shown in Figure 5.6, we found that the Me-CDs can adsorb on the steel surface through the chemisorption of N and O with Fe atoms. Due to its unique chemical structure, quantum dots are considered as amphiphilic. Therefore, the adsorption of CDs helps to reduce the contact between metal surface and corrosive media, that is, to form a hydrophobic protective film, which can effectively inhibit the diffusion of corrosive media on the metal surface [106].

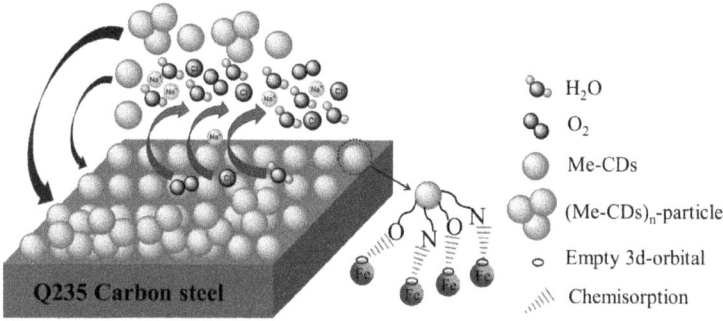

Figure 5.6: Schematic diagram of Me-CDs adsorption on the Q235 carbon steel surface.

5.6 Conclusions and prospect

Quantum dots, as zero-dimensional nanomaterial-based corrosion inhibitor, provide a novel way to develop corrosion inhibitor due to the high-efficient and easily decorated properties. Moreover, zero-, one-, two- and three-dimensional nanomaterials largely enrich the category of corrosion inhibitor. The following limitations can be summarized:

(1) First and foremost, the low yield and high cost are the general shortcoming for nanomaterial application, but the good news is some practical methods have been built for further large-scale preparation. For example, Zhou et al. [107] proposed a micro-zone electrochemical on method for the large-scale fabrication of nitrogen-doped carbon quantum dots, and the utilization rate of raw materials reached 90.18 wt%.

(2) On the other hand, one notable limitation for nanomaterial as corrosion inhibitor is the poor long-term corrosion inhibition performance, where the agglomeration phenomenon for nanomaterials brings huge damage for the corrosion protection process.

(3) The larger molecular weight induces the slower mass transfer rate and complex adsorption behavior, and the adsorption model for nanomaterial corrosion inhibitor has not been unified. The complex reconstruction process makes it difficult to propose the anticorrosion mechanism for nanomaterial as corrosion inhibitor.

In the further study, in view of the abovementioned limitations, the detailed mass transfer process, adsorption model and film formation of quantum dot inhibitors need to be clarified during the corrosion protection process. And some new techniques need to be built to obtain the spatiotemporal information. In a word, the design and application of nanomaterial corrosion inhibitor provide the novel direction in the field of corrosion control.

References

[1]   Liao, B.K., Cen, H.Y., Chen, Z.Y., Guo, X.P., Corrosion behavior of Sn-3.0 Ag-0.5 Cu alloy under chlorine-containing thin electrolyte layers, Corrosion Science, 2018, 143, 347–361.

[2]   Wang, H., Quan, X.D., Zeng, Q., Wu, Y., Liao, B.K., Guo, X.P., Electrochemical corrosion and protection of low-temperature sintered silver nanoparticle paste in NH$_4$Cl solution, Journal of Materials Science: Materials in Electronics, 2021, 32, 13748–13760.

[3]   Kang, L., Zeng, Q., Wang, B., Zeng, J.S., Liao, B.K., Wu, H.X., Cheng, Z., Guo, X.P., Facile fabrication of multi superlyophobic nano soil coated-mesh surface with excellent corrosion resistance for efficient immiscible liquids separation, Separation and Purification Technology, 2022, 284, 120266.

[4]   Liao, B.K., Chen, Z.Y., Qiu, Q.B., Guo, X.P., Inhibitory effect of cetyltrimethylammonium bromide on the electrochemical migration of tin in thin electrolyte layers containing chloride ions, Corrosion Science, 2017, 118, 190–201.

[5]   Liao, B.K., Chen, Z.Y., Qiu, Q.B., Zhang, G.A., Guo, X.P., Effect of citrate ions on the electrochemical migration of tin in thin electrolyte layer containing chloride ions, Corrosion Science, 2016, 112, 393–401.

[6]   Liao, B.K., Wang, H., Xiao, W.P., Cai, Y., Guo, X.P., Recent advances in method of suppressing dendrite formation of tin-based solder alloys, Journal of Materials Science: Materials in Electronics, 2020, 31, 13001–13010.

[7]   Liao, B.K., Wei, L.S., Chen ZY, X., Guo, X.P., Na 2 S-influenced electrochemical migration of tin in a thin electrolyte layer containing chloride ions, RSC Advances, 2017, 7, 15060–15070.

[8]   Zeng, Q., Min, X.H., Luo, Z.G., Dai, H.F., Liao, B.K., In-situ preparation of superhydrophobic Zn-Al layered double hydroxide coatings for corrosion protection of aluminum alloy, Material Letters, 2022, 328, 133077.

[9]   Wan, S., Chen, H.K., Ma, X.Z., Chen, L.J., Kang, L., Liao, B.K., Dong, Z.H., Guo, X.P., Anticorrosive reinforcement of waterborne epoxy coating on Q235 steel using NZ/BNNS nanocomposites, Progress in Organic Coatings, 2021, 159, 106410.

[10] Wan, S., Chen, H.K., Cai, G.Y., Liao, B.K., Guo, X.P., Functionalization of h-BN by the exfoliation and modification of carbon dots for enhancing corrosion resistance of waterborne epoxy coating, Progress in Organic Coatings, 2022, 165, 106757.

[11] Cao, J., Lv, Z., Liao, B.K., Chen, D.P., Tong, W., Zong, Z.F., Li, C., Xiang, T.F., In-situ fabrication of superhydrophobic surface on copper with excellent anti-icing and anti-corrosion properties, Materials Today Communications, 2022, 33, 104633.

[12] Liao, B.K., Luo, Z.G., Wan, S., Chen, L.J., Insight into the anti-corrosion performance of acanthopanax senticosus leaf extract as eco-friendly corrosion inhibitor for carbon steel in acidic medium, Journal of Industrial and Engineering Chemistry, 2022, 117, 238–246.

[13] Liao, B.K., Chen, H.Y., Xiang, T.F., Dai, H.F., Wu, H.X., Wan, S., Guo, X.P., Functionalized nanocomposites as corrosion inhibitors, Functionalized Nanomaterials for Corrosion Mitigation: Synthesis, Characterization, and Applications, 2022, 10, 213–229.

[14] Wan, S., Cen, H.Y., Zhang, T., Liao, B.K., Guo, X.P., Anti-corrosion mechanism of parsley extract and synergistic iodide as novel corrosion inhibitors for carbon steel-Q235 in acidic medium by electrochemical, XPS and DFT methods, Frontiers in Bioengineering and Biotechnology, 2021, 9.

[15] Sharma, S., Kumar, A., Recent advances in metallic corrosion inhibition: A review, Journal of Molecular Liquids, 2021, 322, 114862.

[16] Chaubey, N., Qurashi, A., Chauhan, D.S., Quraishi, M.A., Frontiers and advances in green and sustainable inhibitors for corrosion applications: A critical review, Journal of Molecular Liquids, 2021, 321, 114385.

[17] Gregg, M., Ramachandran, S., Review of corrosion inhibitor developments and testing for offshore oil and gas production systems, Corrosion-Us, 2004.

[18] Tamalmani, K., Husin, H., Review on corrosion inhibitors for oil and gas corrosion issues, Applied Sciences, 2020, 10, 3389.

[19] Pulikkalparambil, H., Siengchin, S., Parameswaranpillai, J., Corrosion protective self-healing epoxy resin coatings based on inhibitor and polymeric healing agents encapsulated in organic and inorganic micro and nanocontainers, Nano-Structures and Nano-Objects, 2018, 16, 381–395.

[20] Fateh, A., Aliofkhazraei, M., Rezvanian, A., Review of corrosive environments for copper and its corrosion inhibitors, Arabian Journal of Chemistry, 2020, 13, 481–544.

[21] Raja, P.B., Ismail, M., Ghoreishiamiri, S., Mirza, J., Ismail, M.C., Kakooei, S., Rahim, A.A., Reviews on corrosion inhibitors: A short view, Chemical Engineering Communications, 2016, 203, 1145–1156.

[22] Goyal, M., Kumar, S., Bahadur, I., Verma, C., Ebenso, E.E., Organic corrosion inhibitors for industrial cleaning of ferrous and non-ferrous metals in acidic solutions: A review, Journal of Molecular Liquids, 2018, 256, 565–573.

[23] Verma, C., Haque, J., Quraishi, M.A., Ebenso, E.E., Aqueous phase environmental friendly organic corrosion inhibitors derived from one step multicomponent reactions: A review, Journal of Molecular Liquids, 2019, 275, 18–40.

[24] Quraishi, M.A., Chauhan, D.S., Ansari, F.A., Development of environmentally benign corrosion inhibitors for organic acid environments for oil-gas industry, Journal of Molecular Liquids, 2021, 329, 115514.

[25] Winkler, D.A., Breedon, M., White, P., Hughes, A.E., Sapper, E.D., Cole, I., Using high throughput experimental data and in silico models to discover alternatives to toxic chromate corrosion inhibitors, Corrosion Science, 2016, 106, 229–235.

[26] Twite, R.L., Bierwagen, G.P., Review of alternatives to chromate for corrosion protection of aluminum aerospace alloys, Progress in Organic Coatings, 1998, 33, 91–100.

[27] Winkler, D.A., Breedon, M., Hughes, A.E., Burden, F.R., Barnard, A.S., Harvey, T.G., Cole, I., Towards chromate-free corrosion inhibitors: Structure–property models for organic alternatives, Green Chemistry, 2014, 16, 3349–3357.

[28] Bahadur, A., Development and evaluation of a low chromate corrosion inhibitor for cooling water systems, Canadian Metallurgical Quarterly, 1998, 37, 459–468.

[29] Gharbi, O., Thomas, S., Smith, C., Birbilis, N., Chromate replacement: What does the future hold?, Npj Materials Degradation, 2018, 2, 12.

[30] Tan, B., Xiang, B., Zhang, S., Qiang, Y., Xu, L., Chen, S., He, J., Papaya leaves extract as a novel eco-friendly corrosion inhibitor for Cu in H2SO4 medium, Journal of Colloid and Interface Science, 2021, 582, 918–931.

[31] Guo, L., Zhang, R., Tan, B., Li, W., Liu, H., Wu, S., Locust Bean Gum as a green and novel corrosion inhibitor for Q235 steel in 0.5 M H2SO4 medium, Journal of Molecular Liquids, 2020, 310, 113239.

[32] Wang, Q., Tan, B., Bao, H., Xie, Y., Mou, Y., Li, P., Chen, D., Shi, Y., Li, X., Yang, W., Evaluation of Ficus tikoua leaves extract as an eco-friendly corrosion inhibitor for carbon steel in HCl media, Bioelectrochemistry, 2019, 128, 49–55.

[33] Udensi, S.C., Ekpe, O.E., Nnanna, L.A., Newbouldia laevis leaves extract as tenable eco-friendly corrosion inhibitor for aluminium alloy AA7075-T7351 in 1 M HCL corrosive environment: Gravimetric, electrochemical and thermodynamic studies, Chemistry Africa, 2020, 3, 303–316.

[34] Verma, C., Ebenso, E.E., Quraishi, M.A., Ionic liquids as green and sustainable corrosion inhibitors for metals and alloys: An overview, Journal of Molecular Liquids, 2017, 233, 403–414.

[35] Kobzar, Y.L., Fatyeyeva, K., Ionic liquids as green and sustainable steel corrosion inhibitors: Recent developments, Chemical Engineering, 2021, 425, 131480.

[36] Uerdingen, M., Treber, C., Balser, M., Schmitt, G., Werner, C., Corrosion behaviour of ionic liquids, Green Chemistry, 2005, 7, 321–325.

[37] Zunita, M., Kevin, Y.J., Ionic liquids as corrosion inhibitor: From research and development to commercialization, Results in Engineering, 2022, 15, 100562.

[38] Kumar, A., Kumar, J., Natural gums as corrosion inhibitor: A review, Materials Today: Proceedings, 2022, 64, 141–146.

[39] Ismail, A., A review of green corrosion inhibitor for mild steel in seawater, Arpn Journal of Engineering and Applied Sciences, 2016, 11, 8710–8714.

[40] Feng, Y.Y., He, J.H., Zhan, Y.L., An, J.B., Tan, B.C., Insight into the anti-corrosion mechanism Veratrum root extract as a green corrosion inhibitor, Journal of Molecular Liquids, 2021, 334, 116110.

[41] Huang, L., Chen, W.Q., Wang, S.S., Zhao, Q., Li, H.J., Wu, J.C., Starch, cellulose and plant extracts as green inhibitors of metal corrosion: A review, Environmental Chemistry Letters, 2022, 20, 3235–3264.

[42] Salleh, S.Z., Yusoff, A.H., Zakaria, S.K., Taib, M.A.A., Seman, A.A., Masri, M.N., Mohamad, M., Mamat, S., Sobri, S.A., Ali, A., Teo, P.T., Plant extracts as green corrosion inhibitor for ferrous metal alloys: A review, Journal of Cleaner Production, 2021, 304, 127030.

[43] Habibiyan, A., Ramezanzadeh, B., Mahdavian, M., Kasaeian, M., Facile size and chemistry-controlled synthesis of mussel-inspired bio-polymers based on Polydopamine Nanospheres: Application as eco-friendly corrosion inhibitors for mild steel against aqueous acidic solution, Journal of Molecular Liquids, 2020, 298, 111974.

[44] Dehghani, A., Bahlakeh, G., Ramezanzadeh, B., Ramezanzadeh, M., Potential role of a novel green eco-friendly inhibitor in corrosion inhibition of mild steel in HCl solution: Detailed macro/micro-scale experimental and computational explorations, Construction and Building Materials, 2020, 245, 118464.

[45] Marzorati, S., Verotta, L., Trasatti, S., Green corrosion inhibitors from natural sources and biomass wastes, Molecules, 2018, 24, 48.

[46] Jain, P., Patidar, B., Bhawsar, J., Potential of nanoparticles as a corrosion inhibitor: A review, Journal of Bio-and Tribo-Corrosion, 2020, 6, 1–12.

[47] Kandasamy, K., Surendhiran, S., Devi, R.C., Khadar, Y.S., Rajasingh, P., Balamurugan, A. Facile and eco-friendly synthesis of CdS quantum dots for enhancing corrosion inhibition of Zn metal plate in various environments, in: AIP Conference Proceedings. AIP Publishing LLC 2022, 2385, 020001.

[48] Palaniappan, N., Cole, I.S., Caballero-Briones, F., Balasubaramanian, K., Lal, C., Praseodymium-decorated graphene oxide as a corrosion inhibitor in acidic media for the magnesium AZ31 alloy, RSC Advances, 2018, 8, 34275–34286.

[49] Zamindar, S., Murmu, M., Murmu, N.C., Banerjee, P., Chemically modified graphene and graphene oxides as corrosion inhibitors, Carbon Allotropes, 2015, 149.

[50] Kavimani, V., Rajesh, R., Rammasamy, D., Selvaraj, N.B., Yang, T., Prabakaran, B., Jothi, S., Electrodeposition of r-GO/SiC nano-composites on magnesium and its corrosion behavior in aqueous electrolyte, Applied Surface Science, 2017, 424, 63–71.

[51] Jiang, L., Dong, Y., Yuan, Y., Zhou, X., Liu, Y., Meng, X., Recent advances of metal–organic frameworks in corrosion protection: From synthesis to applications, Chemical Engineering Journal, 2022, 430, 132823.

[52] Fouda, A.E.A.S., Etaiw, S.E.D.H., El-bendary, M.M., Maher, M.M., Metal-organic frameworks based on silver (I) and nitrogen donors as new corrosion inhibitors for copper in HCl solution, Journal of Molecular Liquids, 2016, 213, 228–234.

[53] Etaiw, A.E.D.H., El-bendary, M.M., Fouda, A.E.A.S., Maher, M.M., A new metal-organic framework based on cadmium thiocyanate and 6-methylequinoline as corrosion inhibitor for copper in 1 M HCl solution, Protection of Metals and Physical Chemistry, 2017, 53, 937–949.

[54] Cui, M.J., Ren, S.M., Xue, Q.J., Zhao, H.C., Wang, L.P., Carbon dots as new eco-friendly and effective corrosion inhibitor, Journal of Alloys and Compounds, 2017, 726, 680–692.

[55] Berdimurodov, E., Verma, D.K., Kholikov, A., Akbarov, K., Guo, L., The recent development of carbon dots as powerful green corrosion inhibitors: A prospective review, Journal of Molecular Liquids, 2022, 349, 118124.

[56] Ye, Y.W., Zhang, D.W., Zou, Y.J., Zhao, H.C., Chen, H., A feasible method to improve the protection ability of metal by functionalized carbon dots as environment-friendly corrosion inhibitor, Journal of Cleaner Production, 2020, 264, 121682.

[57] Karthikeyan, P., Sathishkumar, S., Pandian, K., Mitu, L., Rajavel, R., Novel copper doped Halloysite Nano Tube/silver-poly (pyrrole-co-3, 4-ethylenedioxythiophene) dual layer coatings on low nickel stainless steel for anti-corrosion applications, Journal of Science: Advanced Materials and Devices, 2018, 3, 59–67.

[58] Hsieh, Y.P., Hofmann, M., Chang, K.W., Jhu, J.G., Li, Y.Y., Chen, K.Y., Yang, C.C., Chang, W.S., Chen, L.C.Y., Complete corrosion inhibition through graphene defect passivation, Acs Nano, 2014, 8, 443–448.

[59] Prasai, D., Tuberquia, J.C., Harl, R.R., Jennings, G.K., Bolotin, K.I., Graphene: Corrosion-inhibiting coating, Acs Nano, 2012, 6, 1102–1108.

[60] Prabakar, S.J.R., Hwang, Y.H., Bae, E.G., Lee, D.K., Pyo, M., Graphene oxide as a corrosion inhibitor for the aluminum current collector in lithium ion batteries, Carbon, 2013, 52, 128–136.

[61] Jahdaly, B.A.A., Elsadek, M.F., Ahmed, B.M., Farahat, M.F., Taher, M.M., Khalil, A.M., Outstanding graphene quantum dots from carbon source for biomedical and corrosion inhibition applications: A review, Sustainability-Basel, 2021, 13, 2127.

[62] Verma, C., Alfantazi, A., Quraishi, M.A., Quantum dots as ecofriendly and aqueous phase substitutes of carbon family for traditional corrosion inhibitors: A perspective, Journal of Molecular Liquids, 2021, 343, 117648.

[63] Zhou, Q.Z., Yuan, G.H., Lin, M.J., Wang, P.P., Li, S.J., Tang, J., Lin, J.S., Huang, Y.Y., Zhang, Y., Large-scale electrochemical fabrication of nitrogen-doped carbon quantum dots and their application as corrosion inhibitor for copper, Journal of Materials Science, 2021, 56, 12909–12919.

[64] Lv, J., Fu, L.P., Zeng, B., Tang, M.J., Li, J.B., Synthesis and acidizing corrosion inhibition performance of N-doped carbon quantum dots, Russian Journal of Applied Chemistry, 2019, 92, 848–856.

[65] Xu, X., Wei, H.Y., Liu, M.G., Zhou, L.S., Shen, G.Z., Li, Q., Hussain, G., Yang, F., Fathi, R., Chen, H., Ostrikov, K. (K), Nitrogen-doped carbon quantum dots for effective corrosion inhibition of Q235 steel in concentrated sulphuric acid solution, Materials Today Communications, 2021, 29, 102872.

[66] Zhu, M.Y., Guo, L., He, Z.Y., Marzouki, R., Zhang, R.H., Berdimurodov, E., Insights into the newly synthesized N-doped carbon dots for Q235 steel corrosion retardation in acidizing media: A detailed multidimensional study, Journal of Colloid and Interface Science, 2022, 608, 2039–2049.

[67] Kalajahi, S.T., Rasekh, B., Yazdian, F., Neshati, J., Taghavi, L., Green mitigation of microbial corrosion by copper nanoparticles doped carbon quantum dots nanohybrid, Environmental Science and Pollution Research, 2020, 27, 40537–40551.

[68] Cen, H.Y., Chen, Z.Y., Guo, X.P., N, S co-doped carbon dots as effective corrosion inhibitor for carbon steel in CO_2-saturated 3.5% NaCl solution, Journal of the Taiwan Institute of Chemical Engineers, 2019, 99, 224–238.

[69] Cen, H.Y., Zhang, X., Zhao, L., Chen, Z.Y., Guo, X.P., Carbon dots as effective corrosion inhibitor for 5052 aluminium alloy in 0.1 M HCl solution, Corrosion Science, 2019, 161, 108197.

[70] Zhang, Y., Zhang, S.T., Tan, B.C., Guo, L., Li, H.T., Solvothermal synthesis of functionalized carbon dots from amino acid as an eco-friendly corrosion inhibitor for copper in sulfuric acid solution, Journal of Colloid and Interface Science, 2021, 604, 1–14.

[71] Zhang, Y., Tan, B.C., Zhang, X., Guo, L., Zhang, S.T., Synthesized carbon dots with high N and S content as excellent corrosion inhibitors for copper in sulfuric acid solution, Journal of Molecular Liquids, 2021, 338, 116702.

[72] Saraswat, V., Yadav, M., Improved corrosion resistant performance of mild steel under acid environment by novel carbon dots as green corrosion inhibitor, Colloid Surface A, 2021, 627, 127172.

[73] Saraswat, V., Kumari, R., Yadav, M., Novel carbon dots as efficient green corrosion inhibitor for mild steel in HCl solution: Electrochemical, gravimetric and XPS studies, Journal of Physics and Chemistry of Solids, 2022, 160, 110341.

[74] Liu, Z.X., Ye, Y.W., Chen, H., Corrosion inhibition behavior and mechanism of N-doped carbon dots for metal in acid environment, Journal of Cleaner Production, 2020, 270, 122458.

[75] Luo, J.X., Cheng, X., Zhong, C.F., Chen, X.H., Ye, Y.W., Zhao, H., Chen, H., Effect of reaction parameters on the corrosion inhibition behavior of N-doped carbon dots for metal in 1 M HCl solution, Journal of Molecular Liquids, 2021, 338, 116783.

[76] Zeng, Y.X., Kang, L., Wu, Y., Wan, S., Liao, B.K., Li, N., Guo, X.P., Melamine modified carbon dots as high effective corrosion inhibitor for Q235 carbon steel in neutral 3.5 wt% NaCl solution, Journal of Molecular Liquids, 2022, 349, 118108.

[77] Ren, S.M., Cui, M.J., Chen, X.Y., Mei, S.X., Qiang, Y.J., Comparative study on corrosion inhibition of N doped and N,S codoped carbon dots for carbon steel in strong acidic solution, Journal of Colloid and Interface Science, 2022, 628, 384–397.

[78] Chen, W.X., Hu, C.F., Yang, Y.H., Cui, J.H., Liu, Y.L., Rapid synthesis of carbon dots by hydrothermal treatment of lignin, Materials, 2016, 9, 184.

[79] Zhu, L.L., Shen, D.K., Liu, Q., Luo, K.H., Li, C., Mild acidolysis-assisted hydrothermal carbonization of lignin for simultaneous preparation of green and blue fluorescent carbon quantum dots, ACS Sustainable Chemistry & Engineering, 2022, 10, 9888–9898.

[80] Zhu, L.L., Shen, D.K., Liu, Q., Wu, C.F., Gu, S., Sustainable synthesis of bright green fluorescent carbon quantum dots from lignin for highly sensitive detection of Fe^{3+} ions, Applied Surface Science, 2021, 565, 150526.

[81] Wang, Y., Liu, Y.S., Zhou, J., Yue, J.Q., Xu, M.C., An, B., Ma, C.H., Li, W., Liu, S.X., Hydrothermal synthesis of nitrogen-doped carbon quantum dots from lignin for formaldehyde determination, RSC Advances, 2021, 11, 29178–29185.

[82] Chen, S.L., Wang, G.H., Sui, W.J., Parvez, A.M., Dai, L., Si, C.L., Novel lignin-based phenolic nanosphere supported palladium nanoparticles with highly efficient catalytic performance and good reusability, Industrial Crops and Products, 2020, 145, 112164.

[83] Abdallah, M., Hegazy, M.A., Alfakeer, M., Ahmed, H., Adsorption and inhibition performance of the novel cationic Gemini surfactant as a safe corrosion inhibitor for carbon steel in hydrochloric acid, Green Chemistry Letters and Reviews, 2018, 11, 457–468.

[84] Chafiq, M., Chaouiki, A., Damej, M., Lgaz, H., Salghi, R., Ali, I.H., Benmessaoud, M., Masroor, S., Chung, I.M., Bolaamphiphile-class surfactants as corrosion inhibitor model compounds against acid corrosion of mild steel, Journal of Molecular Liquids, 2020, 309, 113070.

[85] Zuo, X.L., Li, W.P., Luo, W., Zhang, X., Qiang, Y.J., Zhang, J., Li, H., Tan, B.C., Research of *Lilium brownii* leaves extract as a commendable and green inhibitor for X70 steel corrosion in hydrochloric acid, Journal of Molecular Liquids, 2021, 321, 114914.

[86] Adebayo, M., Akande, S., Olorunfemi, A., Ajayi, O., Orege, J., Daniel, E., Equilibrium and thermodynamic characteristics of the corrosion inhibition of mild steel using sweet prayer leaf extract in alkaline medium, Progress in Chemical and Biochemical Research, 2021, 4, 80–91.

[87] Shao, S., Wu, B.B., Wang, P., He, P., Qu, X.P., Investigation on inhibition of ruthenium corrosion by glycine in alkaline sodium hypochlorite based solution, Applied Surface Science, 2020, 506, 44976.

[88] El-Katori, E.E., Al-Mhyawi, S., Assessment of the Bassia muricata extract as a green corrosion inhibitor for aluminum in acidic solution, Green Chemistry Letters and Reviews, 2019, 12, 31–48.

[89] Florez-Frias, E., Barba, V., Lopez-Sesenes, R., Landeros-Martínez, L.L., Los Ríos, J.P.F.D.L., Casales, M., Gonzalez-Rodriguez, J.G., Use of a metallic complex derived from Curcuma longa as green corrosion inhibitor for carbon steel in sulfuric acid, International Journal of Corrosion, 2021.

[90] Samontha, A., Lugsanangarm, K., Corrosion inhibition and adsorption mechanism of eugenol on copper in HCl medium, Protection of Metals and Physical Chemistry, 2019, 55, 187–194.

[91]   Eddy, N.O., Ameh, P.O., Essien, N.B., Experimental and computational chemistry studies on the inhibition of aluminium and mild steel in 0.1 M HCl by 3-nitrobenzoic acid, Journal of Taibah University Medical Sciences, 2018, 12, 545–556.

[92]   Abdel-Gaber, A.M., Rahal, H.T., Beqai, F.T., Eucalyptus leaf extract as a eco-friendly corrosion inhibitor for mild steel in sulfuric and phosphoric acid solutions, International Journal of Industrial Chemistry, 2020, 11, 123–132.

[93]   Hijazi, K., Abdel-Gaber, A., Younes, G., Habchi, R., Comparative study of the effect of an acidic anion on the mild steel corrosion inhibition using Rhus coriaria plant extract and its quercetin component, Portugaliae Electrochimica Acta, 2021, 39, 237–252.

[94]   Saady, A., Ech-Chihbi, E., El-Hajjaji, F., Benhiba, F., Zarrouk, A., Rodi, Y.K., Taleb, M., Biache, A.E.I., Rais, Z., Molecular dynamics, DFT and electrochemical to study the interfacial adsorption behavior of new imidazo [4, 5-b] pyridine derivative as corrosion inhibitor in acid medium, Journal of Applied Electrochemistry, 2021, 51, 245–265.

[95]   Wan, S., Zhang, T., Chen, H.K., Liao, B.K., Guo, X.P., Kapok leaves extract and synergistic iodide as novel effective corrosion inhibitors for Q235 carbon steel in H_2SO_4 medium, Industrial Crops and Products, 2022, 178, 114649.

[96]   Chauhan, D.S., Mouaden, K.E., Quraishi, M.A., Bazzi, L., Aminotriazolethiol-functionalized chitosan as a macromolecule-based bioinspired corrosion inhibitor for surface protection of stainless steel in 3.5% NaCl, International Journal of Biological Macromolecules, 2020, 152, 234–241.

[97]   Chauhan, D.S., Madhan Kumar, A.M., Quraishi, M.A., Hexamethylenediamine functionalized glucose as a new and environmentally benign corrosion inhibitor for copper, Chemical Engineering Research and Design, 2019, 150, 99–115.

[98]   Cui, M.J., Ren, S.M., Zhao, H.C., Wang, L.P., Xue, Q.J., Novel nitrogen doped carbon dots for corrosion inhibition of carbon steel in 1 M HCl solution, Applied Surface Science, 2018, 443, 145–156.

[99]   Cui, M.J., Li, X., Nitrogen and sulfur Co-doped carbon dots as ecofriendly and effective corrosion inhibitors for Q235 carbon steel in 1 M HCl solution, RSC Advances, 2021, 11, 21607–21621.

[100] Ye, Y.W., Yang, D.P., Chen, H., A green and effective corrosion inhibitor of functionalized carbon dots, Journal of Material Science Technology, 2019, 35, 2243–2253.

[101] Yang, D.P., Ye, Y.W., Su, Y., Liu, S., Gong, D.R., Zhao, H.C., Functionalization of citric acid-based carbon dots by imidazole toward novel green corrosion inhibitor for carbon steel, Journal of Cleaner Production, 2019, 229, 180–192.

[102] Cen, H.Y., Cao, J.J., Chen, Z.Y., Functionalized carbon nanotubes as a novel inhibitor to enhance the anticorrosion performance of carbon steel in CO2-saturated NaCl solution, Corrosion Science, 2020, 177, 109011.

[103] Chang, X.J., Henderson, W.M., Bouchard, D.C., Multiwalled carbon nanotube dispersion methods affect their aggregation, deposition, and biomarker response, Environmental Science Technology, 2015, 49, 6645–6653.

[104] Lorite, I., Romero, J.J., Fernandez, J.F., Influence of the nanoparticles agglomeration state in the quantum-confinement effects: Experimental evidences, AIP Advances, 2015, 5, 037105.

[105] Li, J.B., Lv, J., Fu, L.P., Tang, M.J., Wu, X.D., New ecofriendly nitrogen-doped carbon quantum dots as effective corrosion inhibitor for saturated CO_2 3% NaCl solution, Russian Journal of Applied Chemistry, 2020, 93, 380–392.

[106] Xu, X.K., Li, Y.D., Hu, G.Q., Mo, L.Q., Zheng, M.T., Lei, B.F., Zhang, X.J., Hu, C.F., Zhuang, J.L., Liu, Y.L., Surface functional carbon dots: Chemical engineering applications beyond optical properties, Journal of Materials Chemistry C, 2020, 8, 16282–16294.

[107] Zhou, Q.Z., Yuan, G.H., Lin, M.J., Wang, P.P., Li, S.J., Tang, J., Lin, J.S., Huang, Y.Y., Zhang, Y., Large-scale electrochemical fabrication of nitrogen-doped carbon quantum dots and their application as corrosion inhibitor for copper, Journal of Materials Science, 2021, 56, 12909–12919.

Humira Assad, Ashish Kumar*

Chapter 6
Carbon nanotubes (SWCNTs/MWCNTs) and functionalized carbon nanotubes in corrosion prevention

Abstract: Deploying metallic components and gadgets in commercial areas is fraught with danger from corrosion. Strategies for preventing corrosion have undergone a lot of development. Despite extensive research on the topic, no perfect defense has been identified to completely prevent corrosion. Fortunately, techniques to slow down progressive corrosion have been tried and have been extremely successful. These techniques include employing protective coatings and altering the chemistry or structure of the material. To increase the corrosion resistance of composites, research has moved its attention to the integration of innovative materials and structures. To take advantage of the useful corrosion-protective properties of the nanocomposite, nanofillers like fullerene, nanosized carbon allotropes, in general, graphene, graphene oxide (GO) and carbon nanotubes (single- and multiwalled carbon nanotubes (SWCNTs and MWCNTs)) have been added to polymeric matrices. CNTs have become a sought-after and promising filler due to their inert biological nature and remarkable mechanical, electrical and thermal properties. By establishing a passive coating over metallic substances and encouraging sacrificial fortification, notably in Zn-rich polymer (ZRP) coverings, CNTs can seal the voids in metallic substances and polymer-based composites, acting as an anticorrosion filler. In this regard, the chapter attempts to give an insight of how CNT functionalization can improve the corrosion prevention of various metals in various corrosive media, followed by the application of SWCNTs/MWCNTs in protection against corrosion.

Keywords: Corrosion, functionalization, carbon nanotubes, prevention, nanotechnology, nanocomposites

6.1 Introduction

The process of corrosion is when a metal interacts physically and chemically with its environment, changing the metal's characteristics and impairing the functionality of the material, the environment, or a system including both [1–4]. A drop in the system's Gibbs

---
*Corresponding author: Ashish Kumar, NCE, Bihar Engineering University, Department of Science and Technology, Government of Science and Technology, Government of Bihar, Bihar, 803108, India
Humira Assad, Department of Chemistry, School of Chemical Engineering and Physical Sciences, Lovely Professional University, Punjab, India

https://doi.org/10.1515/9783111071756-006

free energy primarily brings on corrosion. The metal consequently exhibits a significant attraction to revert to its original, lower energy oxide state. Significant problems with the economy, security and environment are brought on by corrosion. The price of corrosion globally is projected to be US \$2.5 trillion or approximately 3.4% of world GDP according to a recent report by the National Association of Chemical Engineers [5–8]. Therefore, it may be important to safeguard the substance from this unfavorable action when the integrity of the substance is altered by an external action from the atmosphere. The best method to safeguard metallic surfaces is to introduce coatings. Inorganic, organic and hybrid protective coatings have all been investigated in order to prevent or reduce metallic damage brought on by corrosion processes [9]. Organic coatings do not, however, offer long-lasting corrosion protection and are prone to fault formation. When employed as metal coatings, neat polymers have also demonstrated deprived corrosion fortification and wear characteristics [10]. Thus, matrices have been supplemented with nano-sized carbon nanoparticles to increase the corrosion resistance of polymeric nanocomposites (PNC). Graphene, GO, single-walled carbon nanotubes (SWCNTs) and multiwalled carbon nanotubes (MWCNTs), as well as their chemically modified derivatives, are among the nanoscale carbon allotropes (CA) that are frequently utilized in anticorrosive coating compositions [11]. Typically, the strong hydrophobicity and nanofiller properties of CA make them the perfect anticorrosive materials. Recently, polymers have been used in conducting and nonconducting ways to protect metals against corrosion. Utilizing CA as anticorrosive materials has several benefits, but they also come with some unique difficulties such as uncontrolled dispersion in polymer matrices (PM). CNT-containing PM composites have received more attention than other composites in comparison [12]. The material can be altered by creating a composite or by alloying it. CNTs are a desirable option for corrosion protection applications due to their superior mechanical, morphological and electrical capabilities as well as their chemical inertness. Utilizing CNTs in metal matrix composites (MMCs) presents a number of difficulties including the reduced wettability of C by liquefied metal, the emergence of interfacial reaction yields and related galvanic corrosion [13]. Along with enhancing the mechanical qualities of MMC, CNTs may also improve the performance of electrochemical reactions, heat dissipation, sensing and wear resistance. Several researchers looked into the use of CNT in coatings because of their superior mechanical and lubricating qualities. The use of CNTs in these systems was made possible by simple processing techniques such electrodeposition, electroless deposition, spraying, molecular level mixing and sputtering. Although some extremely good findings have been recorded, the improvements in the CNT-containing composites' characteristics are often not as striking as one may have anticipated. This is mostly attributable to adhesion, the most fascinating problem that CNTs face and the inability to achieve a uniform dispersion of filler inside the matrix. As-produced CNTs have a high surface energy and hence have a tendency to entangle and form agglomerates, which results in a substantially less uniform dispersion and worsens the characteristics of the composite [14]. In order to address this issue, the surface of the CNTs is typically altered using covalent (such as acid treatment) or noncovalent techniques (such as surfactants). In a recent study, Dlugon et al. [15] created a CNT coating on the surface of Ti by electroplating, and they found that the incorporation of CNT increased the deterioration of Ti. Although no explanation for this behavior was offered, it is quite unlikely to obtain a dense and uniform CNT covering; as a result, some parts of the Ti substrate are exposed to the corrosive solution, which will encourage crevice corrosion. A flawed coating will not only improve corrosion resistance but will also actually make it worse. Moreover, Khun et al. [16] investigated the impact of CNT concentration on epoxy coating's ability to prevent corrosion on AA2024-T3. According to electrochemical impedance spectroscopy (EIS) data, by boosting the coating's CNT content, the epoxy's pore resistance would increase, indicating that ionic conductivity is hampered. Additionally, CPE1 shows that employing CNT reduces the penetration of the electrolyte. Besides, R_{ct} shows an increase in substrate resistance to anodic dissolution. Thus, it was concluded that adding CNT to the epoxy coating enhanced its ability to prevent corrosion. This enhancement was attributable to the removal of MWNTs from epoxy's ionic conduction pathways. In other words, CNTs decreased the epoxy coating's through-porosity. The prospective of CNTs as a persuasive corrosion inhibitor (CI) will be examined using potentiodynamic polarization (PDD) and EIS in this chapter. Surface modification and functionalization of C nanotubes (SWCNTs/MWCNTs) for corrosion protectiveness of several metals will also be covered. Finally, a critical appraisal and the prospects for further developments are covered. Finally, a critical appraisal and the prospects for further developments are covered.

6.2 Functionalization of CNTs

Owing to their exceptional electrical, mechanical and thermal capabilities, CNTs have received a lot of interest since they were discovered by Iijima in 1991 [17, 18]. A SWCNT, which is the most common type of CNT, is a sheet of graphite one atom thick that has been trundled into a tube with a diameter of 1 nm. If there are surplus or manifold graphene tubes surrounding the hub of an SWCNT, this is identified as a MWCNT, as shown in Figure 6.1. Nanometer fractions and tens of nanometers are used to measure diameter. Lengths can reach several centimeters, and both of their ends are typically covered in a structure resembling fullerenes [19].

The remarkable qualities of CNTs are thought to have ushered in a new age in the world of materials, particularly in the development of conductive polymers and CNT-centered NC for corrosion prevention. It is a possible reinforcement for polymers for operational and structural implementations due to their excellent characteristics. However, the addition of CNTs could also change the way that polymers protect against corrosion and their tribological characteristics [21, 22]. Hu et al. [23] discovered that the MWCNTs could plug the micropores and imperfections on the exterior of the Pb–Sn-electroplated coverings to operate as an effective physical barrier increasing the corrosion resistance. MWCNTs can reduce the friction coefficient and wear loss pace

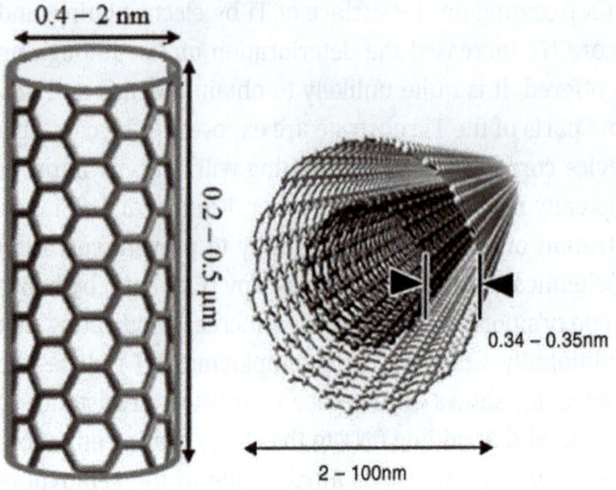

Figure 6.1: Schematic representation of SWCNTs (left) and MWCNTs (right) [20].

of BMI resin according to Liu et al. [24]. However, owing to their outsized precise surface area, weak interfacial interaction with PM, and van der Waals (VW) and π–π interactions, it is challenging to scatter CNTs in polymers and have negative effects on the characteristics of CNTs/PNC. The distribution of CNTs and their suitability with PM have been improved using a variety of techniques, which can be segmented into covalent and noncovalent techniques (mechanical stirring, sonication, centrifugal mingling, usage of dispersants and block copolymers like surfactants, and sonication) [25–28] as illustrated in Figure 6.2.

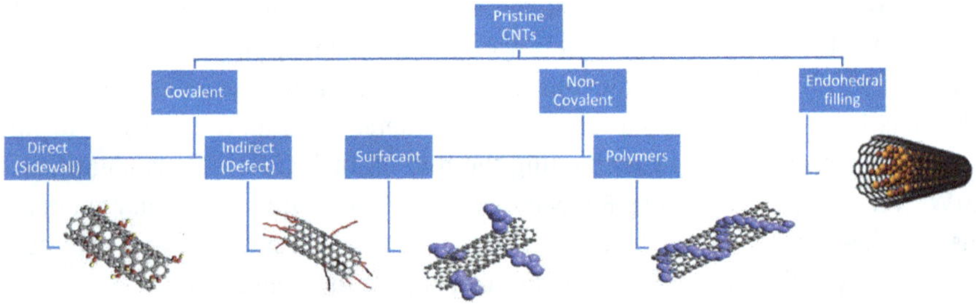

Figure 6.2: Functionalization methods for carbon nanotubes [20].

Typically, covalent connection can be used to functionalize one or even more specific locations on CA [29]. Due to their exceptional electrical conductivity, CNTs accelerate the process of electron transfer, which increases corrosion in aquatic environments. Therefore, although having numerous advantages, the use of CNTs as CIs, especially in the aqueous phase, is rather limited. They can, however, be covalently functionalized to gain desired anticorrosive characteristics. SWCNTs and MWCNTs can be covalently functionalized via a variety of chemical processes [30]. Covalent functionalization (CF) of both SWCNTs and MWCNTs results in the physiochemical modifications listed below:

- Widening of the dispersion
- Improvement in the capacity for interfacial adhesion
- A reduction in hydrophobicity
- A decline in agglomeration capacity
- A loss of mechanical strength
- The electrical conductivity is decreasing

As a result, while numerous flaws are unavoidably produced on the CNT sidewalls, covalent techniques can impart beneficial functional groups into the CNT surface. In some extreme situations, the CF procedure can fragment CNTs into smaller pieces, particularly when combined with the destructive ultrasonication procedure, which can cause a significant loss in the mechanical characteristics of CNTs [31]. Functional molecules can operate as dispersants in physical methods, adhering to the lateral walls of CNTs through VW interactions and aggregating between CNTs and polymer chains with aromatic rings (devoid of abolishing the construction of CNTs). To preserve the ideal framework and attributes of CNTs, it is more desirable than the covalent method. Importantly, covalently functionalized CNTs (both single- and multiwalled) can be further derivatized (as required) relying on the marginal substituents existing in their molecular configurations [32].

In contrast, noncovalent functionalization (NCF) of CA is mostly accomplished through hydrophobic contacts, electrostatic interactions, VW intermolecular force of attraction and π–π stacking [33, 34]. NCF, which preserves the majority of the CA's features, including their electrical and thermal properties, obviously presents an interesting procedure for functionalizing CA. Numerous studies have used π–π stacking – stacking to NCF CA, particularly CNTs (SWCNTs and MWCNTs). For this kind of contact, the π–$\bar{e}$ cloud of the aromatic ring(s) (such as pyrene and perylene) typically cooperates with the π–$\bar{e}$ cloud of the CA. Aromatic compounds and CA interact by a process known as "π–π-stacking," whereas communications between CA (CNTs) and macromolecules, surfactants and ionic liquids are known as hydrophobic interactions. CA dispersion in both aqueous and organic solvents is primarily achieved through the use of these interactions. NCF of CA can be accomplished via H bonding in addition to π–π stacking, hydrophobic and electrostatic interactions [29].

6.3 Corrosion prevention by CNTs

Allotropes of carbon, particularly CNTs, are receiving a lot of interest these days for their potential application as CIs. The widespread use of CNT and its derivatives is explained by a variety of attractive characteristics such as outstanding mechanical resilience, elevated thermal and chemical confrontation, extreme surface-to-volume ratio and elevated dispensability as well as outstanding proficiency to relate with the metal outer layers [35, 36]. Besides, being used as an anticorrosive, CNTs and their derivatives are efficaciously employed as chemical transformation catalysts, surfaces for the elimination of hazardous heavy metal ions, electrochemical detectors and polymer-centered composites [36]. Due to their high heat and electrical conductivity, CNTs are predicted to accelerate corrosion when used in their purest form. CNTs are often produced using a variety of techniques such as chemical vapor deposition, the thermal fabrication method and arc discharge evaporation [37]. CNT derivatives, particularly polymer composites, are being extensively researched as anticorrosive supplies for ferrous and nonferrous alloys due to their wide range of intriguing features. Composites can decrease the gap size and raise the coating's compactness to stop the absorption of corrosive fluids owing to the interwoven network framework of CNTs in the coating. Furthermore, the greater prospective of the CNT composite can encourage the passivation of metallic specimen and produce a shielding cover [38]. In addition, the greater potential of the CNT composite can help metals passivate and produce a protective barrier [39]. Nevertheless, the microgalvanic connections between the cathodic CNTs and the anodic matrix in the alloy may cause the addition of CNTs to hasten corrosion. Therefore, it is still unclear whether CNTs contribute to corrosion inhibition or enhance it. CNTs have, however, infrequently been noted to act as a CI in solutions. Baghalha et al. [40] investigated the CNT nanofluids' ability to suppress the corrosion of copper in sodium dodecyl sulfate solution, and they found that, despite the lack of adsorption process analysis, the layer of adsorbed CNTs had a notable impact on the corrosion process. Ionita and Pruna [41] used molecular mechanics and molecular dynamics methodologies to anticipate the mechanical characteristics of "(PPy)/polyaminobenzene sulfonic acid-functionalized SWCNTs (CNT-PABS) and PPy/carboxylic acid-functionalized SWCNTs (CNT-CA)." For CS (OL 48–50) alloys in 3.5% sodium chloride, PPy film, CNT-PABS and CNT-CA were also used as corrosion-resistant composite material. The PDP, scanning electron microscope (SEM) and transmission electron microscope (TEM) methodologies were utilized to evaluate the anticorrosive properties of different compositions. The results demonstrate that CNTPABS and CNT-CA are distributed consistently across composite materials and are more effective at preventing corrosion than pure PPy. Results also indicated that the order in which the different formulations inhibited corrosion was: PPy < CNT-CA < CNT-PABS. Yousefi et al. [42] used a variety of methodologies to investigate the performance of inhibition of NCF of CNTs using decomposable Gemini surfactants on MS outer layer in 2 M HCl solution. The findings show that NCF of CNTs with ester-encompassing surfactants exhibited more suitable inhibition features at extreme surfactant dosages, with the best inhibition efficiency (IE_{Eis} = 93%) conveyed for BT (2.5 mM) overhanging nanotubes because of further dispersing capability. Surface observations were also used to confirm the effectiveness of the CNTs' NCF in preventing corrosion on MS.

For quantum chemical calculations, density functional theory was used, and a strong correlation between experimental and hypothetical statistics was found [42].

Another study has looked into the anticorrosive impact of PPy/MWCNT NC for 304 stainless steel deterioration in 3% NaCl mixture [43]. A polyaniline/f-MWCNT NC was created and used by Kumar et al. [44] as an anticorrosive covering for mild steel in a neutral NaCl mixture (3.5%). In order to characterize the fabricated nanocomposite, Raman, ATR-IR and FE-SEM investigations were performed. MWCNT showed outstanding scattering in the PM of PANI according to FESEM research. In the PANI PM, the f-CNTs were securely retained in place and evenly dispersed. The diameter of the f-CNTs dramatically increases in the occurrence of PANI, according to subsequent FESEM research, which is explained by the PANI polymer coating of the f-CNT surface. The hydrophobicity of the nanocomposite was evaluated using a contact angle study. According to the PDP investigation, the inclusion of PAN and PCNT greatly reduces the corrosion current density, with PCNT having the biggest impact. According to an EIS investigation, the existence of PANI and PCNT raises the value of R_{ct} (charge transfer resistance).

Because of their capacity to adhere on the metal's exterior, PANI and PCNT both function as interface-category CIs, which is how they prevent corrosion. The presence of PCNT3 caused the value of R_{ct} to grow the most. R_{ct} values increased in the following order: PCNT3 > PNCT2 > PCNT1 > PANI. For the value of corrosion current densities, only inverse order was seen. It has also been looked into if the inclusion of PANI-MWCNT composite-based alkyd covering for MS degradation in 1 M HCl solution increases charge transfer resistance (R_{ct}) values [45]. For anticorrosive coverings based on alkyd, PANI/alkyd and PANI-MWCNT/alkyd, respectively, the R_{ct} values of 151, 2,172 and 10,300 Ωcm^2 were calculated. Calabrese Luigi investigated a sol–gel N-propyl-trimethoxy-silane coating coated with different amounts of MWCNTs to increase the impedance of Al to corrosion. Drop casting was utilized to operate the NC coating to the AA6061 Al alloy substrate as shown in Figure 6.3. The morphological examination showed that 0.4 wt% CNT produced a homogeneous sol–gel covering [46].

According to the study, adding CNT to the base covering considerably increased the corrosion fortification of the Al alloy in an NaCl 3.5 wt% electrolyte (seawater). A drop in the corrosion current of at least two orders of magnitude was seen in PDD curves, which also showed remarkable results, as shown in Figure 6.4. Best outcomes were seen on AS3-CNT6 specimens, which had a breaking potential of 0.620 V/AgAgCl$_{sat}$ and a passivation current density of almost 1.0×10^{-5} mA/cm^2 [46].

By in situ oxidative polymerizing $C_6H_5NH_2$ on the outer layer of MWCNTs, Qiu et al. [47] created the CNT-polyaniline (c-PANI). Studies revealed that the diameter of the nanobrushes is influenced by the mass ratio of aniline to MWCNTs. The synthetic nanobrushes have outstanding anticorrosive action against epoxy resin with an amine cure. The nanobrush coating has good anticorrosive properties in neutral, low pH, and basic pH conditions. Potentiodynamic examination demonstrates that coatings based on c-PANI composites shown comparatively good protective efficiency when compared to coatings based on PANI in all investigated media (various pH). According to estimates, c-PANI is more electrochemically reactive than PANI in a range of surfaces due to the

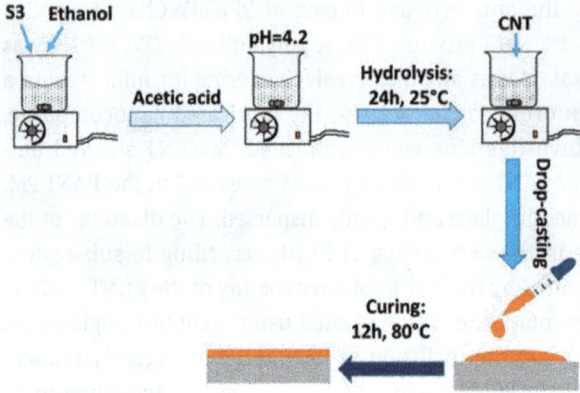

Figure 6.3: Schematic representation of coating process deposition of carbon nanotubes (CNT) [46].

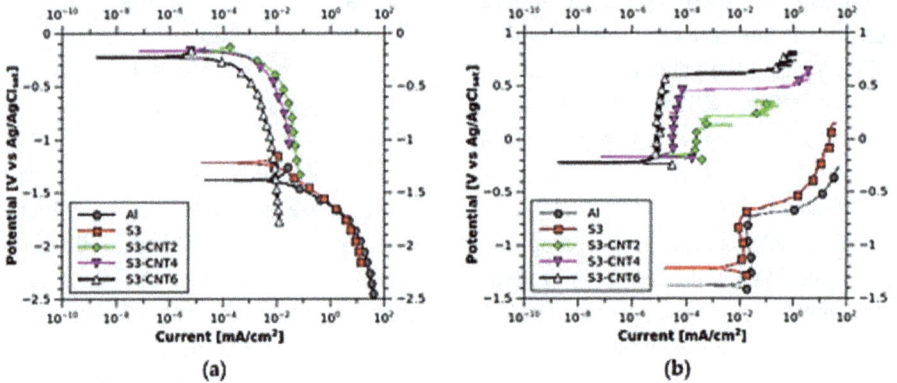

Figure 6.4: Potentiodynamic polarization curves for the (a) cathodic and (b) anodic branches of CNTs in an NaCl 3.5 wt% solution [46].

interaction between polyaniline chains and MWCNTs via hydrogen bonding and stacking interactions. According to additional analysis, c-PANI befits effectual in acidic environments by passivating the CS outer layer, whereas in neutral and high pH environments, c-PANI develops effectual by building a physical shield. Other investigations have also looked into the PANI-CNT-based nanocomposites' ability to inhibit [48]. Graphitic filamentous nano-carbon-aligned carbon thin layer (GFNACTL) for Cu in simulated sea H_2O, conductive polyurethane-MWCNTs composites for SS in 3% NaCl, and other CNT-based nanocomposites have similar corrosion prevention properties [49, 50].

PDA/CNT nanocomposite coating's impact on carbon steel's (CS) ability to reduce corrosion was studied by Hong et al. [51]. The corrosion resistance of CS specimen was explored using EIS techniques. Electrochemical studies revealed that, as shown in Figure 6.5, corrosion impedance of CS covered with PDA and PDA/CNTs was significantly more than that of naked CS.

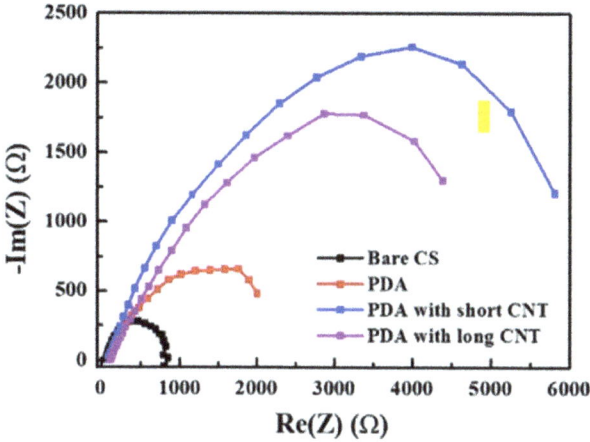

Figure 6.5: Nyquist plot of PDA/CNT NC coating for CS corrosion protection [51].

Two processes were identified as the cause of this corrosion mitigation. According to the progress of both the cathodic and anodic curvatures in the PD diagram, PDA can reduce corrosion rates by creating a lean, thick coating on the outer layer. Second, the cathodic partial process was suppressed by CNTs because of O_2 adherence at the defect sites of the CNTs, according to cathodic curves from the PD analysis. This explains why the short-CNT coating has better anticorrosion properties than the long-CNT coating, as illustrated in Figure 6.6.

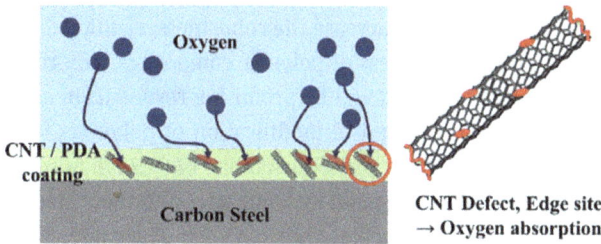

Figure 6.6: Diagram illustrating nanocomposite coating's role in reducing corrosion [51].

Additionally, Cen et al. [52] created functionalized carbon nanotubes (FCNTs) using a hydrothermal process, which they then utilized as a CI for CS in an NaCl mixture that is CO_2-saturated. The results showed that FCNTs can successfully prevent CS corrosion, and at a dosage of 100 mg/L, their IE is around 90%. Through the use of functionalized groups, FCNTs can attach to CS and form the hydrophobic coating that prevents corrosive fluids from diffusing across the metal surface.

In order to prevent the entrance of detrimental moieties such as H_2O and O_2, CNTs consistently disperse and plug the micropores prevalent on the surface coatings.

Anticorrosive coatings made of polymers perform better and last longer when CNTs are present. For aqueous electrolytes, CNTs serve as anticorrosive barriers due to their hydrophobic nature. The bulk of the time, CNT-based nanocomposites are found to act as interface-type CIs because they adsorb at the metal/electrolyte interface and become effective there. The value of R_{ct} increases in direct proportion to the adsorption of CNT-based nanocomposites on metallic surfaces. The metallic surface's active sites may also be blocked by this kind of adsorption. These locations typically cause corrosive dissolution. Hence, the value of corrosion current density is lessened when CNT-based nanocomposites are present in anticorrosive coatings (I_{corr}). Several surface investigation techniques, such as scanning electron microscopy, TEM, Fourier transformed infrared spectroscopy, X-ray photoelectron spectroscopy, X-ray diffraction, energy-dispersive X-ray analysis and ultraviolet–visible spectroscopy (UV-vis) can be used to access the adsorption of CNT-based NC on metals.

6.4 Conclusion and outlook

As anticorrosive coatings, CAs, particularly CNTs and their derivatives, are frequently utilized due to their extensive surface area, filler content and hydrophobicity. In several polymer-based coatings, they are commonly utilized as nanofiller. CNTs have ushered in a new era of polymer composites for use in a variety of structural applications. CNTs are added to polymer composite constructions with natural fiber. It goes without saying that they fix and fill the rapture coating structures. Outstanding performance has helped to justify the quick rise in publications and has encouraged the creation of engineering goods. CNTs have been found to greatly increase the robustness, rigidity, durability, electrical conductivity and thermal steadiness of polymer composites. According to the literature discussed in this chapter, there should be promising results from additional investigation on the combination and chemical modification of polymers with CNTs for a new generation of superior composites. The repeatability of CTNs in implications on a huge scale, and their homogenous dispersion in composites and their interfacial interactions, are pertinent technological hurdles that must be overcome. They easily aggregate, which negatively affects their dispersion qualities and ability to stop corrosion. These substances' dispensability can be upgraded by properly functionalizing them with organic complexes. They can also be physically dispersed by using techniques like magnetic stirring, ultrasonic mixing, ball milling and shear emulsification. However, there are comparatively fewer studies on CA as aqueous phase CIs. With the aid of organic materials, attempts are being undertaken to chemically alter those. Modified CA has exceptional solubility in the polar electrolytes and is consequently effective as CIs in the aq. medium. Therefore, future study should emphasize specifically on creating and fabricating derivatives of soluble CA. Utilizing the synergistic effect, that is, by incorporation few inorganic salts such as KI and $ZnCl_2$, their level of protection can be increased even higher. The usage of covalently and noncovalently altered CA in aqueous phase electrolytes should be investigated, particularly in the existence of inorganic salts, in light of the foregoing. Additionally, further research is needed to determine the protection, dependability and sturdiness of these CNT-based composites. There is no uncertainty that CNT hybrids with natural fiber polymer composites should encourage groundbreaking exploration projects and support commercial advancements, but supplementary basic research is required to better understand how the components interact in this exploding, novel class of polymer composite materials.

References

[1]   Nazeer, A.A., Madkour, M., Potential use of smart coatings for corrosion protection of metals and alloys: A review, Journal of Molecular Liquids, 2018, 253, 11–22.
[2]   Assad, H., Kumar, A., Understanding functional group effect on corrosion inhibition efficiency of selected organic compounds, Journal of Molecular Liquids, 2021, 344, 117755.
[3]   Ganjoo, R., et al., Experimental and theoretical study of sodium cocoyl glycinate as corrosion inhibitor for mild steel in hydrochloric acid medium, Journal of Molecular Liquids, 2022, 364, 119988.
[4]   Sharma, S., et al., Multidimensional analysis for corrosion inhibition by Isoxsuprine on mild steel in acidic environment: Experimental and computational approach, Journal of Molecular Liquids, 2022, 357, 119129.
[5]   Shekari, E., Khan, F., Ahmed, S., Economic risk analysis of pitting corrosion in process facilities, International Journal of Pressure Vessels and Piping, 2017, 157, 51–62.
[6]   Umoren, S.A., Solomon, M.M., Saji, V.S., Corrosion Inhibitors for Sour Oilfield Environment (H_2S Corrosion), Corrosion Inhibitors in the Oil and Gas Industry, 2020, 229–254.
[7]   Thakur, A., et al., Computational and experimental studies on the efficiency of *Sonchus arvensis* as green corrosion inhibitor for mild steel in 0.5 M HCl solution, Materials Today: Proceedings, 2022, 66, 609–621.
[8]   Assad, H., Ganjoo, R., Sharma, S., A theoretical insight to understand the structures and dynamics of thiazole derivatives, in Journal of Physics: Conference Series, IOP Publishing, 2022.
[9]   Pour-Ali, S., Dehghanian, C., Kosari, A., Corrosion protection of the reinforcing steels in chloride-laden concrete environment through epoxy/polyaniline–camphorsulfonate nanocomposite coating, Corrosion Science, 2015, 90, 239–247.
[10]   Kausar, A., A review of filled and pristine polycarbonate blends and their applications, Journal of Plastic Film & Sheeting, 2018, 34(1), 60–97.
[11]   Assad, H., Fatma, I., Kumar, A., An overview of the application of graphene-based materials in anticorrosive coatings, Materials Letters, 2022, 133287.
[12]   Bakshi, S.R., Lahiri, D., Agarwal, A., Carbon nanotube reinforced metal matrix composites-a review, International Materials Reviews, 2010, 55(1), 41–64.
[13]   Tjong, S.C., Recent progress in the development and properties of novel metal matrix nanocomposites reinforced with carbon nanotubes and graphene nanosheets, Materials Science and Engineering: R: Reports, 2013, 74(10), 281–350.
[14]   Daneshvar-Fatah, F., Nasirpouri, F., A study on electrodeposition of Ni-noncovalnetly treated carbon nanotubes nanocomposite coatings with desirable mechanical and anti-corrosion properties, Surface and Coatings Technology, 2014, 248, 63–73.

[15] Dlugon, E., et al., Carbon nanotube-based coatings on titanium, Bulletin of Materials Science, 2015, 38(5), 1339–1344.

[16] Khun, N., Troconis, B.R., Frankel, G., Effects of carbon nanotube content on adhesion strength and wear and corrosion resistance of epoxy composite coatings on AA2024-T3, Progress in Organic Coatings, 2014, 77(1), 72–80.

[17] Sanli, A., et al., Piezoresistive performance characterization of strain sensitive multi-walled carbon nanotube-epoxy nanocomposites, Sensors and Actuators A: Physical, 2017, 254, 61–68.

[18] Ahmadi, M., et al., Synergistic effect of MWCNTs functionalization on interfacial and mechanical properties of multi-scale UHMWPE fibre reinforced epoxy composites, Composites Science and Technology, 2016, 134, 1–11.

[19] Ghosh, P., Kumar, K., Chaudhary, N., Influence of ultrasonic dual mixing on thermal and tensile properties of MWCNTs-epoxy composite, Composites Part B: Engineering, 2015, 77, 139–144.

[20] Aslam, M.M.-A., et al., Functionalized carbon nanotubes (Cnts) for water and wastewater treatment: Preparation to application, Sustainability, 2021, 13(10), 5717.

[21] Shen, W., et al., Multiwall carbon nanotubes-reinforced epoxy hybrid coatings with high electrical conductivity and corrosion resistance prepared via electrostatic spraying, Progress in Organic Coatings, 2016, 90, 139–146.

[22] Kumar, A., et al., Thermo-mechanical and anti-corrosive properties of MWCNT/epoxy nanocomposite fabricated by innovative dispersion technique, Composites Part B: Engineering, 2017, 113, 291–299.

[23] Hu, Z., Jie, X., Lu, G., Corrosion resistance of Pb–Sn composite coatings reinforced by carbon nanotubes, Journal of Coatings Technology and Research, 2010, 7(6), 809–814.

[24] Liu, L., et al., The effects of the variations of carbon nanotubes on the micro-tribological behavior of carbon nanotubes/bismaleimide nanocomposite, Composites Part A: Applied Science and Manufacturing, 2007, 38(9), 1957–1964.

[25] Cha, J., et al., Functionalization of carbon nanotubes for fabrication of CNT/epoxy nanocomposites, Materials & Design, 2016, 95, 1–8.

[26] Huang, J., et al., Enhancement of lignocellulose-carbon nanotubes composites by lignocellulose grafting, Carbohydrate Polymers, 2017, 160, 115–122.

[27] Fatma, I., et al., Interaction behavior of N-Lauroyl Sarcosine Sodium Salt (NLSS) and Benzethonium Chloride (BC) in Aqueous Human Serum Albumin (HSA) at different temperatures: A volumetric and acoustic study, Journal of Chemical & Engineering Data, 2022, 67(11), 3385–3399.

[28] Fatma, I., et al., Influence of HSA on micellization of NLSS and BC: An experimental-theoretical approach of its binding characteristics, Journal of Molecular Liquids, 2022, 367, 120532.

[29] Verma, C., et al., Recent advancements in corrosion inhibitor systems through carbon allotropes: Past, present, and future, Nano Select, 2021, 2(12), 2237–2255.

[30] Wu, H.-C., et al., Chemistry of carbon nanotubes in biomedical applications, Journal of Materials Chemistry, 2010, 20(6), 1036–1052.

[31] Ma, P.-C., et al., Dispersion and functionalization of carbon nanotubes for polymer-based nanocomposites: A review, Composites Part A: Applied Science and Manufacturing, 2010, 41(10), 1345–1367.

[32] Karim, M.R., et al., Proton conductors produced by chemical modifications of carbon allotropes, perovskites and metal organic frameworks, Journal of Materials Chemistry A, 2017, 5(16), 7243–7256.

[33] Marcia, M., Hirsch, A., Hauke, F., Perylene-based non-covalent functionalization of 2D materials, FlatChem, 2017, 1, 89–103.

[34] Shown, I., Ganguly, A., Non-covalent functionalization of CVD-grown graphene with Au nanoparticles for electrochemical sensing application, Journal of Nanostructure in Chemistry, 2016, 6(4), 281–288.

[35] Peng, Y.G., et al., Preparation of poly (m-phenylenediamine)/ZnO composites and their photocatalytic activities for degradation of CI acid red 249 under UV and visible light irradiations, Environmental Progress & Sustainable Energy, 2014, 33(1), 123–130.

[36] Zare, E.N., Lakouraj, M.M., Ramezani, A., Efficient sorption of Pb (II) from an aqueous solution using a poly (aniline-co-3-aminobenzoic acid)-based magnetic core–shell nanocomposite, New Journal of Chemistry, 2016, 40(3), 2521–2529.

[37] Salvetat, J.-P., et al., Mechanical properties of carbon nanotubes, Applied Physics A, 1999, 69(3), 255–260.

[38] Cho, S., Chiu, T.-M., Castaneda, H., Electrical and electrochemical behavior of a zinc-rich epoxy coating system with carbon nanotubes as a diode-like material, Electrochimica Acta, 2019, 316, 189–201.

[39] Rui, M., Jiang, Y., Zhu, A., Sub-micron calcium carbonate as a template for the preparation of dendrite-like PANI/CNT nanocomposites and its corrosion protection properties, Chemical Engineering Journal, 2020, 385, 123396.

[40] Baghalha, M., Kamal-Ahmadi, M., Copper corrosion in sodium dodecyl sulphate solutions and carbon nanotube nanofluids: A modified Koutecky–Levich equation to model the agitation effect, Corrosion Science, 2011, 53(12), 4241–4247.

[41] Ioniţă, M., Prună, A., Polypyrrole/carbon nanotube composites: Molecular modeling and experimental investigation as anti-corrosive coating, Progress in Organic Coatings, 2011, 72(4), 647–652.

[42] Yousefi, A., et al., An experimental and theoretical study of biodegradable Gemini surfactants and surfactant/carbon nanotubes (CNTs) mixtures as new corrosion inhibitor, Journal of Bio-and Tribo-Corrosion, 2019, 5(4), 1–15.

[43] Ganash, A., Electrochemical synthesis and corrosion behaviour of polypyrrole and polypyrrole/carbon nanotube nanocomposite films, Journal of Composite Materials, 2014, 48(18), 2215–2225.

[44] Kumar, A.M., Gasem, Z.M., In situ electrochemical synthesis of polyaniline/f-MWCNT nanocomposite coatings on mild steel for corrosion protection in 3.5% NaCl solution, Progress in Organic Coatings, 2015, 78, 387–394.

[45] Farag, A.A., et al., Influence of polyaniline/multiwalled carbon nanotube composites on alkyd coatings against the corrosion of carbon steel alloy, Corrosion Reviews, 2017, 35(2), 85–94.

[46] Calabrese, L., Khaskoussi, A., Proverbio, E., Wettability and anti-corrosion performances of carbon nanotube-silane composite coatings, Fibers, 2020, 8(9), 57.

[47] Qiu, G., Zhu, A., Zhang, C., Hierarchically structured carbon nanotube–polyaniline nanobrushes for corrosion protection over a wide pH range, RSC Advances, 2017, 7(56), 35330–35339.

[48] Rajyalakshmi, T., et al., Enhanced charge transport and corrosion protection properties of polyaniline–carbon nanotube composite coatings on mild steel, Journal of Electronic Materials, 2020, 49(1), 341–352.

[49] Jeong, N., et al., One-pot large-area synthesis of graphitic filamentous nanocarbon-aligned carbon thin layer/carbon nanotube forest hybrid thin films and their corrosion behaviors in simulated seawater condition, Chemical Engineering Journal, 2017, 314, 69–79.

[50] Wei, H., et al., Anticorrosive conductive polyurethane multiwalled carbon nanotube nanocomposites, Journal of Materials Chemistry A, 2013, 1(36), 10805–10813.

[51] Hong, M.-S., et al., Polydopamine/carbon nanotube nanocomposite coating for corrosion resistance, Journal of Materiomics, 2020, 6(1), 158–166.

[52] Cen, H., Cao, J., Chen, Z., Functionalized carbon nanotubes as a novel inhibitor to enhance the anticorrosion performance of carbon steel in CO2-saturated NaCl solution, Corrosion Science, 2020, 177, 109011.

Dheeraj Singh Chauhan
Chapter 7
Graphene (Gr)/graphene oxide (GO) and functionalized Gr/GO in corrosion prevention

Abstract: Graphene and graphene oxide (GO), are conventionally used in the development of anticorrosion coatings. While single/multi-layered graphene films are used as a barrier, the GO is used as a filler material inside the epoxy-based coatings. The poor aqueous dispersibility of graphene restricts its application as a direct use in aqueous corrosion inhibitor development. On the other hand, GO, due to the presence of multiple O-containing surface functional groups, shows immense potential to undergo chemical modification. This has led to the development of effective corrosion inhibitors from modified GO. This chapter discusses the potentiality of graphene films and GO in the anticorrosion coatings as well as in the corrosion inhibitors for aqueous acidic, and in saline environments. Some of the different methods of preparation of anticorrosion coatings, and surface functionalization of GO are outlined. A brief review of literature on the chemically functionalized GO in coatings and in corrosion inhibition is presented with mechanism details for interested readers and researchers.

Keywords: Corrosion inhibitor, coatings, graphene, graphene oxide, chemical functionalization

7.1 Introduction

Corrosion is the destruction or loss of material upon interaction with the surrounding medium [1–3]. It is a global menace posing an immense impact on the world economy, estimated at US $2.5 trillion, equivalent to 3.4% of the world's GDP. There are different forms of corrosion such as uniform corrosion, pit formation, crevice formation, galvanic, intergranular corrosion, selective leaching, stress and erosion–corrosion [4, 5]. Different industrial practices such as acid-pickling, acid cleaning and oil well acidizing lead to the direct contact of the metal/alloy surface with concentrated acids causing severe corrosion [6–9]. Cooling systems, boilers, heat exchangers, water treatment plants, etc., expose the metallic materials' surface directly to saline environments, which cause corrosion attack on the metallic structures [10–13]. There are different

Dheeraj Singh Chauhan, Modern National Chemicals, Second Industrial City, Dammam 31421, Saudi Arabia, e-mail: dheeraj.chauhan.rs.apc@itbhu.ac.in

https://doi.org/10.1515/9783111071756-007

methods used to counter corrosion such as cathodic protection, corrosion inhibitors, anticorrosion coatings and the use of corrosion resistant alloys. Several inorganic/organic molecules, polymers and nanomaterials have been employed as effective coating agents and as corrosion inhibitors [4, 5, 14–16].

Graphene (Gr) consists of a two-dimensional hexagonal network of sp^2-hybridized carbon atoms [17–20]. The attractive chemical, thermal stability and impermeability make Gr a good candidate and as a barrier agent for anticorrosive coatings [21–23]. Extensive research is available on single/multilayered Gr and chemically functionalized Gr [24–26]. Gr finds applications for corrosion-inhibiting coatings in the fields such as aircraft components [21]. The important members of the Gr family include (i) single-layer Gr (SLG), (ii) multilayer Gr (MLG), (iii) Gr oxide (GO), (iv) reduced Gr oxide (rGO) and (v) Gr platelets. Other lesser known carbon allotropes include Gr (with both benzene and acetylene bonds) and graphdiyne (with both benzene and diacetylene bonds) [18]. Figure 7.1 shows the structures of different Gr derivatives.

The emergence of the Gr and its derivatives has paved a pathway to significant development within a very short span of time in the diverse areas of human life [27]. The applications include catalysis, drug delivery, medical science, electronics, batteries, the energy industry, construction materials, water treatment and military goods [18, 28, 29]. One of the potential areas of the application of Gr and Gr oxide (GO) is the corrosion protection of metals and alloys [30]. Some review articles have appeared concerning

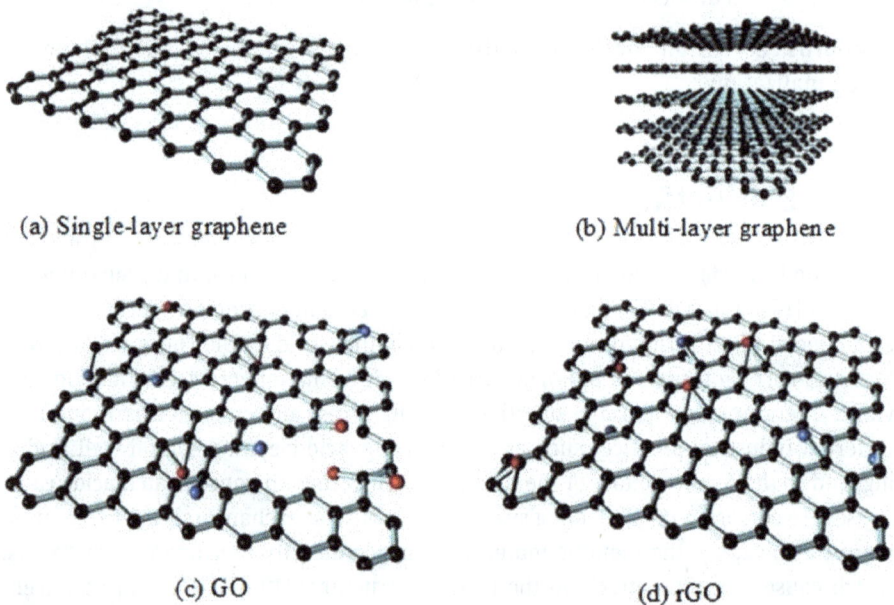

(a) Single-layer graphene

(b) Multi-layer graphene

(c) GO

(d) rGO

Figure 7.1: Structure of graphene and its derivatives: (a) single-layer graphene; (b) multilayer graphene; (c) GO; (d) rGO (reproduced with permission from Ref. [33] © Copyright Elsevier 2020).

the application of Gr and GO in the development of anticorrosion coatings and in corrosion inhibition [31, 32]. In the present chapter, different methods for the synthesis of Gr and the derivatives of Gr, the different routes of chemical functionalization of the nanomaterials, and their application in anticorrosion coatings and corrosion inhibitors are explained. Some drawbacks have been identified, and the scope of further research in this area has been outlined.

7.2 Importance of Gr and GO in corrosion protection

The nanomaterials of the Gr family and GO have conventionally been employed for the development of anticorrosion coatings. Due to its excellent hydrophobic nature, impermeability, chemical stability and excellent barrier properties, Gr is used for application in anticorrosion coatings. Extensive research work is reported on the use of single- and multilayered graphene coatings for the protection of metals and alloys. The chemical inertness and thermal stability of the Gr films have purported their application in anticorrosive coatings. A wide variety of chemical functionalization methods have been reported in the literature on the development of Gr-based coatings involving the covalent and noncovalent functionalization of the Gr and GO. In recent years, Gr-based coatings for various ferrous and nonferrous materials have attracted considerable attention. Organic heterocycles are commonly used as corrosion inhibitors for metals and alloys in diverse corrosive environments [34–36]. However, their synthesis and purification requires a cumbersome protocol and the involvement of skilled organic chemists. Moreover, these inhibitors often are toxic toward environment. To counter this issue, there has been a considerable research in the development of nanomaterial-based corrosion inhibitors. While the nanomaterial-based coatings have been used in the neutral and saline environments primarily, the nanomaterial-based corrosion inhibitors are reported for acidic media such as for the acid pickling of steels and even in the oil-well acidizing. A noteworthy observation is that these inhibitors have been noticed to provide a considerable extent of protection to the underlying metal surface at lower inhibitor dosages compared to conventional organic corrosion inhibitors. In addition, these inhibitors have shown to perform efficiently even at elevated temperatures. Furthermore, the researchers have also examined the effect of several synergistic agents to obtain the improvement in the adsorption and corrosion inhibition behavior of these inhibitors.

7.3 Anticorrosive coatings based on graphene

Coatings are commonly employed to improve the surface properties of a target substrate, e.g., wettability and adhesion, and to impart corrosion protection [14]. A considerable number of nanomaterials such as metal oxides, noble metals and carbon-based nanomaterials have been used as fillers or film-forming agents for anticorrosion coatings. The Gr exhibits stability, impermeability against wear and tear, antioxidant behavior. Gr possesses excellent transparency, thereby protecting the underlying metal surface from the surrounding environment. Because of these properties, Gr finds applications in anticorrosion coatings [32, 37–39]. Several methods are used for the preparation of coatings such as chemical vapor deposition (CVD), dip coating, spin coating, layer-by-layer self-assembly, sol–gel approach, spray coating, in situ polymerization and electrophoretic deposition.

7.3.1 Single-layered graphene coating

Graphene possesses chemical inertness, stability up to 400 °C, and can be grown on a meter-scale and mechanically transferred to desired surfaces. Both the single- and multilayered Gr films are optically transparent and do not disturb the optical properties of the underlying metallic substrate. CVD is the most commonly adopted method for preparation [40]. Other methods include different chemical, mechanical and electrical/electrochemical coating. Arc discharge method can produce single, bi-, and few-layered G films in the range of 100 nm to a few micrometer size [41, 42]. Gr sheet acts as an inert barrier blocking the corrosive species. The metallic surface can be protected from air oxidation even after heating at 200 °C up to 4 h. A single Gr sheet protected a US penny from 30% H_2O_2 for 2 min (Figure 7.2(a)) [43]. Thin films of Gr have shown anticorrosion effects and surface passivation. A GO/poly(ethylene imine) (PEI) coating was investigated for oxygen barrier effect [44]. It is noteworthy to mention that the protective coating consisting of pure G suffers from poor quality and low coverage of surface. The sp^2 hybridized, the thinness and structural planarity of the hexagonal rings of the Gr present an impregnable barrier against substrate-environment interaction. However, despite the barrier action of the Gr film, the defects present on the film can aggravate the corrosion phenomenon as shown in Figure 7.2(b). The CVD-grown protective coating consisting of pure Gr suffers from poor quality and low surface coverage. Corrosion can proceed after the damage of the film, rendering the exposed surface of the metal amenable to destruction. In addition, some issues are experienced in the scale-up of synthesis and commercialization of the method. Here it should be noted that for the prolonged protection effect, a strong Gr–metal interaction plays a crucial role, preventing rapid intercalation of the oxidizing species at the Gr/metal interface and diminishing the oxidation of the underlying substrate [45].

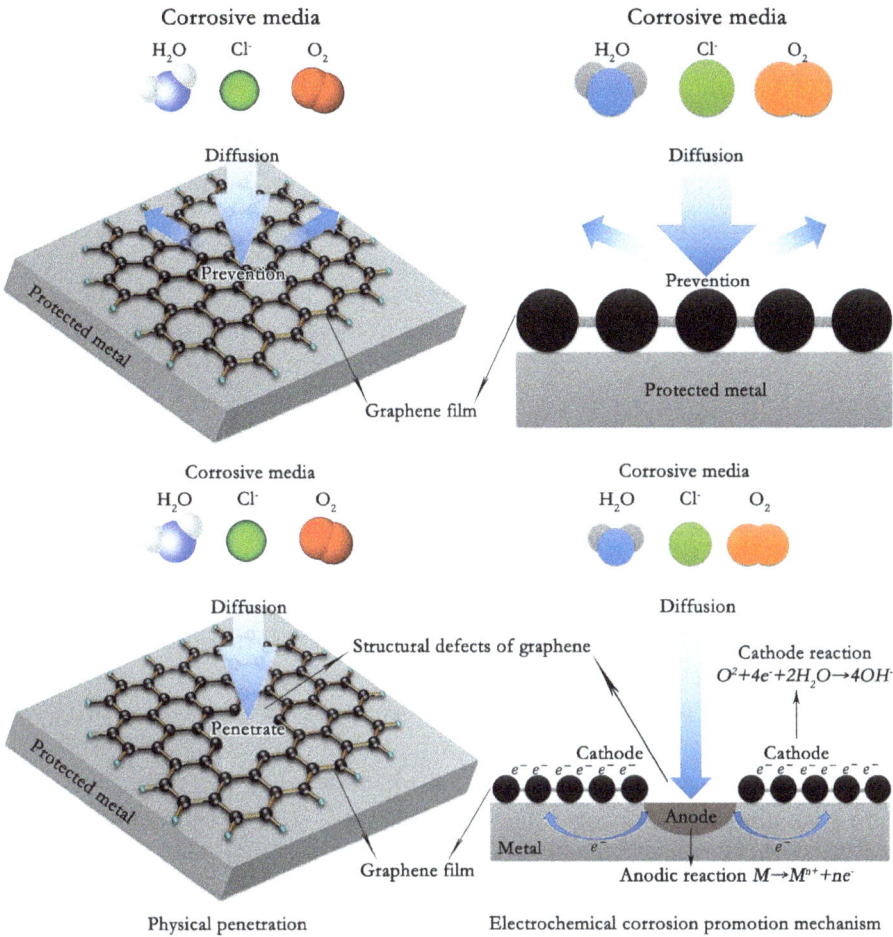

Figure 7.2: (a) Barrier action of Gr film; (b) corrosion promotion through surface defects on Gr film (reproduced with permission from Ref. [21] © Copyright Elsevier 2018).

7.3.2 Multilayered graphene coating

Considering the limitations of the anticorrosion coatings based on SLG, a thicker multilayer Gr (MLG) coatings have been studied to gain improvement in the barrier and corrosion protection. In a study, Cu corrosion rate was reduced twice when they transferred up to three Gr layers on the Cu substrate due to extremely fast mass transport (m/s) [46]. Methanol and ethanol were used as precursors to developing large-area, high-quality Gr nanosheets on Cu foils via the CVD technique [47]. Raman et al. [48] prepared a Gr coating on Cu substrate that increased the resistance of the metallic substrate by 1.5 times. Using atomic layer deposition (ALD) technique, Gr layer defects were passivated, and the corrosion rate of Cu was reduced by over 100 times. Prasai et al. [49] achieved a considerable decrease in the corrosion rate of about 1.6 and 4 times when SLG was prepared on Cu substrate and then transferred onto a Ni substrate. A CVD-grown Gr coating was studied for Cu in 3.5% NaCl, which offered protection to the Cu substrate for a short span of time [50]. In another study, it was shown that upon immersion in NaCl, the Gr layers that were exposed behaved as sites for Fe oxidation [51]. The high mechanical stability of Gr considerably eliminated metal loss and resulted in an oxidation-based increase in mass. On the Ni surface, CVD-grown Gr showed remarkable corrosion protection [52]. On the surface of Inconel 625, a multilayered Gr coating was applied that decreased the corrosion current by two orders in magnitude [53]. Gr coatings prepared on Cu and Al surfaces were studied for anticorrosion effect in 0.1 M NaCl using EIS analysis after 40 days of exposure time. Excellent barrier action and decreased corrosion and dissolution of Cu in 0.1 M NaCl were observed [54]. In situ grown Gr films on carbon steel surface were prepared and Ni atoms were introduced via laser technique leading to the formation of Ni-Fe alloy catalyst [55]. Such films of Gr showed better anticorrosion behavior than Gr's transferred films. On the SUS304 stainless steel surface, Gr films were grown via the CVD technique [56]. This led to a high-performance and eco-friendly anticorrosion coating for the bipolar plate of the PEM fuel cells. Stoot et al. [57] performed a comparison of SS, SS with Ni seed layer, and SS with Ni seed layer coated by MLG film as bipolar plates for PEM fuel cells. The performance of Gr films was much better compared to just the seed layer of Ni toward steel surface protection. In another study, Gr/Zn-containing coatings were prepared, and anticorrosion behavior and water permeation dynamics were studied in simulated seawater [58]. Micro-Raman spectroscopy supported the barrier action of the Gr coatings. A CVD-grown Gr film improved the biocompatible nature and provided anticorrosion behavior to NiTi alloy, suggesting the application of the Gr coatings in the biomedical domain [59]. Gr coatings were also applied to improve the corrosion protection of Au-coated Cu substrates for application in electrical connections [60]. A CVD-deposited multilayer Gr/Ni (MLGr/Ni) coating on a highly oriented pyrolytic graphite electrode afforded an understanding of the mechanism of hydrogen protection as well as anticorrosion behavior [61]. Using metallic substrates Gr films were prepared at low temperatures and utilized for anticorrosion performance [62]. Up to 10 times, a decrease in the corrosion rates was observed on the Gr-grown steel surface. High-power laser beam irradiation technique was used at room temperature to grow Gr coatings on bulk Ni substrate and applied for anticorrosion performance [63]. In situ-grown Gr film afforded considerable protection to the underlying metallic substrate because of strong interaction and remarkable barrier action. A comparison of as-grown and transferred Gr films was carried out on the basis of adhesion properties and electronic, mechanical and frictional performance [64]. A MLG was deposited on Cu surface and provided significant corrosion protection in a chloride solution for prolonged duration ($\sim$400 h) [65]. Antioxidation behavior of rGO-coated multilayers on Fe and Cu foils was studied by transferring from SiO_2 substrate [66]. The films provided excellent diffusion barrier against gas permeation. Table 7.1 provides Gr-based coatings, with preparation techniques, metal surface and corrosive environment.

Table 7.1: Gr-based coatings for corrosion protection [67].

Preparation technique of Gr coating	Metal/medium	References
CVD	Cu/Ni alloy/30% H_2O_2	[43]
CVD	Ni/condensed water vapor	[45]
Atomic layer deposition	Cu/0.1 M Na_2SO_4	[46]
CVD	Cu/0.1 M NaCl	[48]
CVD	Cu/0.1 M Na_2SO_4	[49]
Acetone-derived	Cu/3–3.5% NaCl	[50]
CVD	Ni/0.1 M NaCl	[52]
Low temperature CVD	Inconel 625 alloy/5% HCl	[53]
CVD on Cu, transfer on Al by Cu etching	Cu, Al/0.1 M NaCl	[54]
Laser irradiation	Ni/Fe/3.5% NaCl	[55]
CVD	Stainless steel/3.5% NaCl	[56]
CVD	Stainless steel/3.5% NaCl	[57]
CVD	NiTi alloy/0.9% NaCl	[59]
CVD	Ni/0.5 M H_2SO_4	[61]
CVD	Stainless steel, Cu, Ni and CuNi/5% sea saltwater	[62]
Laser irradiation	Ni/0.5 M HCl	[63]
CVD	Cu/0.1 M NaCl	[65]
CVD	Cu/3.5% NaCl	[68]
CVD	Al alloy/3.5% NaCl	[69]

7.3.3 Gr composite-based coating

In addition to the use of SLG and MLG, the application of GO and rGO dispersed in a coating matrix is extensively reported to improve the barrier action. These nanomaterials are dispersed in a coating and improve the barrier action of the coatings. This combines the strong adhesion behavior of Gr as filler particles and the thin-film-forming characteristic of the matrix, which improves the coating efficiency. Physical dispersions, chemical functionalization and nanoparticle-modified Gr surfaces are some of the commonly used approaches to obtain a uniform dispersion of Gr in the matrix. A functionalized MLG (FMLG) was studied for improvement in adhesion and corrosion resistance of Zn-rich epoxy coating [70]. EIS and salt spray tests revealed that coatings containing FMLG improved the cathodic protection and barrier properties. However, these coatings pose some issues of difficulty in obtaining control over the stacking of Gr nanoplatelets in the coating matrix, leading to complexity in diffusion pathways of the molecules through the coating [71]. Polymers used in the hybrid coatings can play roles in (a) electrical insulation of the Gr from the metallic substrate to reduce or avoid galvanic corrosion, (b) providing adhesion for the coating on a metallic substrate and (c) facilitating ease of transfer of CVD grown Gr on various metallic substrates. The dispersion of Gr within the coating matrix can be accomplished via using physical dispersion, chemical functionalization and surface modification using nanoparticles.

A CTAB-wrapped film of Gr nanoflakes and poly(vinyl chloride) was prepared via simple solution-blending, drop-casting and annealing route to obtain 58% increase in Young's modulus and provided 130% increase in tensile strength [72]. Interactions of Gr and poly(2-butylamine) were utilized to obtain stable dispersions of Gr in organic media [73]. A small portion of this composite in the epoxy coating improved the corrosion protection and wear resistance. Modification of Gr with polymers has been reported to provide coatings with high conductivity and good barrier action [74–78]. Such coatings are still vulnerable to corrosive attack due to the presence of defects in the Gr film. In such conditions, the potential difference between Gr and the metal, the electrical connection of Gr and the metal, and the aggressive electrolyte can be optimized. In such cases, the use of graphene oxide (GO) can facilitate the bonding to the suitable modifiers leading to corrosion protection [79]. A trilaminar structure composed of an electrodeposited Fe–W amorphous alloy layer was prepared using GO cross-linked with silane [80]. Excellent hydrophobic behavior was observed as revealed from the water angle studies. GO-reinforced polyurethane nanocomposite coating was studied for anticorrosion action resulting in an improvement in the hydrophobic nature and corrosion protection [81]. A mechanistic investigation on the improvement in the cathodic corrosion protection using Zn-modified Gr coating was carried out on the steel surface [82]. The ZRE coating in the presence of Gr particles led to a uniform activation of Zn particles and also the superior percolation of Gr nanosheets. Gu et al. [83] reported stable dispersion of Gr in aqueous medium via a carboxy aniline trimer derivative as a stabilizer, in which the commercially obtained Gr was dispersed in aqueous medium in high concentrations of >1 mg/mL. A multifunctional anticorrosion coating was obtained by dispersing Gr into epoxy for protection of Al 2024-T3 aircraft aluminium [84]. Boric acid-induced cross-linked poly(vinyl alcohol-co-ethylene)/GO film was prepared via spray-coating on stainless steel substrate and improved gas barrier and anticorrosion performance was obtained [85]. An epoxy coating was applied on the steel surface, which was pretreated by a silane-coating containing amino and isocyanate silane-functionalized GO nanosheets [86]. The covalent bond formation with the top epoxy coating led to an excellent barrier performance. Using 3-aminopropyltrimethoxysilane (3-APTMS), Al_2O_3 was anchored to the GO surface to afford anticorrosion potential to steel surface [87]. A mixture of organofunctional silane precursors was utilized to prepare the composite matrix employing GO as a filler material, wherein the silanized GO presented a physical barrier to the corrosive species [38]. Surface modification of Gr with poly(methyl methacrylate) provided hydrophobic nature leading to an improvement in the corrosion protection efficiency from 37.90% to 99.53% [88]. 3-APTMS was used to functionalize the Gr to afford an environment-friendly anticorrosion coating [89].

Nanometallic oxides such as TiO_2, Al_2O_3, $CaCO_3$, SiO_2 and $FeO.Fe_2O_3$ have been used to perform surface decoration of Gr to improve dispersibility in the coating matrix and to lower the structural defects. Such modifiers increase the interlayer spacing, lower the agglomeration and promote corrosion resistance of the coating [32, 87, 90–92]. 1-d MWCNTs were used to inhibit the agglomeration of Gr nanosheets [93]. TiO_2 nanoparticles were used for the decoration of Gr via 3-APTES via composite formation resulting in high dispersibility in both aqueous and organic media [94]. GO decorated using Fe_3O_4 was utilized in an epoxy coating resulting in an improvement in the microhardness of the epoxy coating by 71.8% in comparison to a pure epoxy coating at the same loading content [90]. rGO/zinc-rich epoxy composite coatings were designed to improve the corrosion protection of carbon steel substrates [95]. The rGO nanosheets played dual role by forming an impermeable barrier and by allowing a superior electrical conductivity between Zn particles and Fe surface. MWCNTs were dispersed in an urushiol formaldehyde (UFP) polymer via in situ condensation polymerization and the modified GO/MWCNT/UFP composite coatings were prepared using the solution-blending method. The composite exhibited excellent corrosion protection as revealed by electrochemical tests, alkali-resistance, hardness, adhesion and the protection efficiency attained up to 99.7% [96]. The synergism of the functionalized GO and CNT was studied on the anticorrosion potential of the epoxy coating resulting in the superior corrosion protection behavior of the hybrid coatings [97]. Table 7.2 lists some of the coatings prepared from Gr composites.

Table 7.2: List of some graphene composite-based coatings for corrosion protection [67].

Graphene-based nanostructure	Metal/environment	Reference
Poly(2-butylaniline)-functionalized graphene	Q235 steel/3.5% NaCl	[73]
rGO@3-aminopropyl triethoxysilane	Cu/3.5% NaCl	[79]
Graphene-reinforced zinc-rich nanocomposite	Carbon steel/3.5% NaCl	[82]
Carboxylated aniline trimer-stabilized graphene	Carbon steel/3.5% NaCl	[83]
Graphene/epoxy coating	Al 2024-T3/3.5% NaCl	[84]
Poly(vinyl alcohol-*co*-ethylene)/GO	Stainless steel/3.5% NaCl	[85]
Amino and isocyanate silane functionalized GO	Steel/3.5% NaCl	[86]
GO-Al_2O_3 hybrid	Steel/3.5% NaCl	[87]
PMMA nanocomposite embedded with graphene	Steel/3.5% NaCl	[88]
GO-Fe_3O_4 hybrid@polydopamine + KH550	Steel/3.5% NaCl	[90]
Graphene@SiO_2 microsheets	Cu/3.5% NaCl	[98]
Polystyrene (PS)/modified-GO	Steel/3.5% NaCl	[99]
Octadecylamine/graphene oxide/maleic anhydride-grafted polypropylene	Steel/3.5% NaCl	[100]
Fluorographene/epoxy	Cu/3.5% NaCl	[101]

7.4 Synthesis and application of modified GO-based corrosion inhibitors

In addition to the application of GO in anticorrosion coatings, in recent years, several articles have appeared in the literature describing the use of chemically modified GO [25, 29, 102] in corrosion inhibition. While the anticorrosion coatings based on GO have been primarily utilized in saline environments, the GO-based corrosion inhibitors have been studied in saline as well as strongly acidic environments as discussed in the following section. Most of the corrosion inhibition studies are available on mild steel in 1 M HCl solution. For modified GO, the studies have been reported for 1 M HCl as well as a highly aggressive solution of 15% HCl. We have herein covered some of the modification approaches and a brief overview of the literature on the use of modified GO in corrosion inhibition. Figure 7.3 shows different strategies used for the chemical functionalization of GO surface.

7.4.1 Modification using aromatic compounds

GO surface was chemically functionalized using o-phenylenediamine to afford diaminobenzene-GO (DAB-GO) and by o-phenylenediamine and $NaNO_2$ together to afford aminoazobenzene-GO (AAB-GO) [103]. Thus-prepared modified GO was investigated as a corrosion inhibitor for mild steel in 1 M HCl solution using electrochemical studies and surface examination. The increment in the charge transfer resistance (R_{ct}) and decrease in the corrosion current density (i_{corr}) due to the addition of AAB-GO and DAB-GO to the corrosive electrolyte supported the adsorption and protection performance of the GO derivatives onto the mild steel substrate. High efficiencies were afforded at a very low inhibitor dosage. A mixed-type corrosion inhibition behavior with cathodic predominance was observed. XPS studies indicated a metal-inhibitor interaction. DFT-based computational examinations supported the superior performance of AAB-GO in comparison to DAB-GO.

In another study, GO was chemically functionalized via p-aminophenol [104]. p-Aminophenol (PAP)-modified GO investigated for mild steel in 1 M HCl via gravimetric and electrochemical tests. PAP-GO provided high corrosion protection with >92% efficiency, which was greater compared to PAP alone (81.23%) at 25 mg/L. A mixed-type behavior was noted with cathodic predominance. Considering the fact that heterocyclic compounds have been established for effective corrosion inhibition performance, GO surface was functionalized using heterocyclic compounds [15, 16]. For this purpose, GO was functionalized using pyridine to afford diazopyridine (DAZP-GO) and diaminopyridine (DAMP-GO) [105]. GO surfaces modified using pyridine derivatives DAMP and DAZP were also studied for mild steel in 1 M HCl solution [105]. Electrochemical analyses

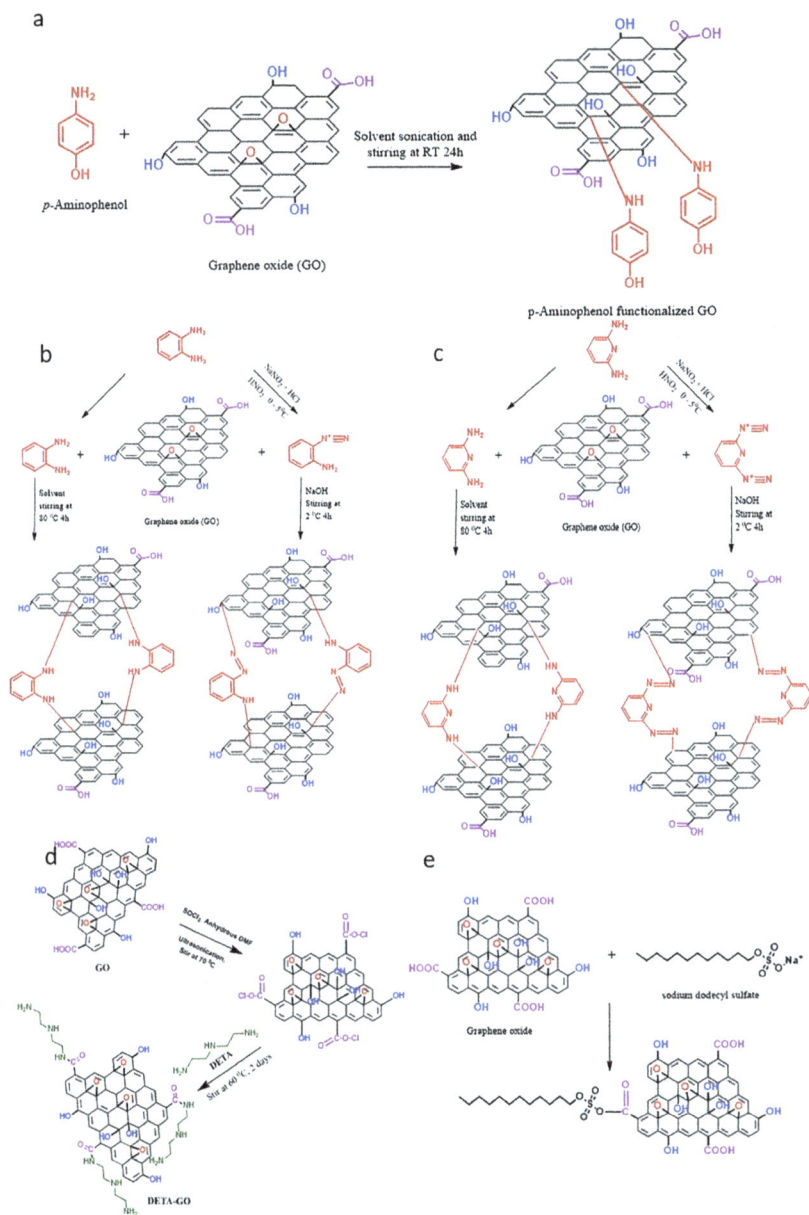

Figure 7.3: Various strategies used for chemical functionalization of GO surface, using (a) p-aminophenol (PAP-GO); (b) diaminopyridine-GO (DAMP-GO) and diazopyridine-GO (DAZP-GO); (c) diaminobenzene-GO (DAB-GO) and aminoazobenzene-GO (AZB-GO); (d) diethylenetriazmine-GO (DETA-GO); (e) sodium dodecyl sulfate-GO (SDS-GO) (reproduced with permission from Ref. [67] © Copyright Elsevier 2020).

revealed a remarkable corrosion inhibition behavior with >95% protection for both GO at 25 mg/L. SEM and AFM studies supported the formation of a smooth inhibitor film on the metallic substrate. XPS indicated the metal-inhibitor interaction and supported formation of inhibitor film.

7.4.2 Modification using surfactants

Surfactants facilitate the solubility and dispersion of nanomaterials and for the conventionally used organic heterocycle-based corrosion inhibitors. Several surfactants as such have been utilized as corrosion inhibitors. Therefore, the surface of GO was modified using two surfactants, namely sodium dodecyl sulfate (SDS) and sodium propyl sulfate (SPS), and studied as corrosion inhibitors for mild steel in 1 M HCl solution [106]. The modification was performed to study the influence of alkyl chain length of the surfactant on the corrosion inhibition performance [106]. It was observed that SDS with a longer alkyl chain showed superior protection performance compared to SPS.

7.4.3 Modification using amines

GO was chemically functionalized using diethylenetriamine (bis(2-aminoethyl)amine) (DETA-GO) and studied for mild steel in 1 M HCl [107]. High efficiency of 93.44% was obtained at a low dose of 25 mg/L. The electrochemical studies were further complemented with DFT-based theoretical studies. In another study, the same inhibitor was analyzed on the surface of carbon steel in 15% HCl solution using weight loss studies and electrochemical measurements [108]. High efficiency at a very low dose of 50 mg/L high efficiencies (>90%) was observed. Weight loss tests were performed at 65 °C for 6 h without and in the presence of 5 mM KI to explore synergism with I^- ions. The introduction of KI led to a high protection performance of >95%. PDP studies revealed a mixed type of inhibition with cathodic predominance. Surface analyses via contact angle and AFM measurements supported the inhibitor adsorption. GO grafted with dopamine was studied as an inhibitor for carbon steel in 15% HCl solution [109]. The weight loss tests showed 80.69% inhibition efficiency at 5 mg/L concentration of the inhibitor. EIS revealed a charge transfer-controlled behavior with PDP showing a mixed-type inhibitor performance. The adsorption of the inhibitor on the metallic surface was in accordance with Langmuir isotherm.

7.4.4 Modification using polymers

Polymers have been widely acknowledged as corrosion inhibitors due to their large molecular weights that can afford a significant coverage of the metallic substrate [110–112].

Further, the polymeric backbone also contains a large number of functional groups that can promote adsorption and protection behavior. β-Cyclodextrin (β-CD)-modified GO was studied as a cost-effective and environment-friendly corrosion inhibitor for steel surfaces in 15% HCl [113]. Results of gravimetric and electrochemical investigations on X60 carbon steel in 15% HCl solution provided a moderate efficiency of 68.61% at 15% v/v concentration. SEM-EDX and FTIR studies supported the adsorption of the CD-GO on the steel surface and the formation of a protective film. The UV-vis studies of the test solution in the presence of inhibitor evidenced the formation of a complex between CD-GO and Fe^{2+} ions. A couple of studies reported the functionalization of the GO surface using PEI as corrosion inhibitor using a similar procedure as that used for diethylenetriamine (DETA) [114, 115]. PEI-GO was studied for corrosion inhibition of carbon steel in 15% HCl without and in the presence of KI at elevated temperatures. At 65 °C after 6 h immersion, >95% efficiency was obtained. PDP revealed a mixed-type nature, with EIS studies showing a charge-transfer controlled behavior. 3-D profilometry showed a considerable lowering in the surface roughness of the steel surface in the presence of the adsorbed corrosion inhibitor. The PEI-GO was also studied for Cu surface in 0.5 M HCl solution via electrochemical analyses [115]. PDP study revealed a mixed mode of corrosion inhibition with cathodic predominance. Cyclic voltammetry examination evidenced the suppression effect of inhibitor on the oxidation–reduction processes of Cu dissolution. Computational studies supported the adsorption and protective action of PEI-GO on metal surface. Researchers also studied the surface decoration of GO using melamine formaldehyde and urea formaldehyde polymers to introduce N, S and O heteroatoms (NSP-GO) [116]. For this purpose, urea (source of N), ammonium sulfate (source of S) and ammonium phosphate (source of P) were used. The corrosion inhibition performance was studied for mild steel in 3.5% NaCl. A comparison of the influence of N, S, P atoms separately and N, S and P together was carried out on the inhibition behavior. A high protection efficiency of 100% was achieved at 500 mg/L. A nanocomposite was prepared using poly(*o*-anthranilic acid), GO and functionalized MWNT (f-MWCNT) and studied as a corrosion inhibitor for stainless steel 316 (SS316) well in 2 M HCl acid solution [117]. High efficiency of 91.35% was obtained at 70 mg/L concentration of the nanocomposite. PDP analyses revealed a mixed-type behavior while the EIS studies showed a charge transfer-controlled phenomenon.

7.4.5 Other modifications

To study the effect of N and S atoms, the GO surface was modified using 2-aminoethanol and 2-mecaptoethanol to obtain GON and GOS, respectively [118]. The obtained inhibitors were used to protect mild steel in 1 M HCl. High corrosion protection of >95% was observed at a low dose of 6 mg/L for both the studied inhibitors. A slight decrease in the inhibition was noted with an increase in the testing temperature. Thermodynamic parameters revealed the physical adsorption of inhibitors on the metal surface. Graphene oxide quantum dots (GOQDs) were produced via electrochemical technique and studied as corrosion inhibitors for Q235 steel in 1 M HCl [119]. The structure of the GOQDs was ascertained using FTIR and XPS analyses. A high inhibition performance was observed in the temperature range of 15–45 °C at a quite low dose of 52.5 mg/L. EIS tests indicated a charge-transfer controlled behavior at room temperature that showed the appearance of inductive loops at elevated temperatures. Table 7.3 lists the performance of different chemically modified GO as corrosion inhibitors.

Table 7.3: Performance of various chemically modified GO-based corrosion inhibitors.

Name	Metal surface/medium	Inhibition efficiency/ concentration	Reference
Aminoazobenzene-GO Diaminobenzene-GO	Mild steel/1 M HCl	96.80/25 mg/L 95.20/25 mg/L	[103]
p-Aminophenol-GO	Mild steel/1 M HCl	92.86/25 mg/L	[104]
Diaminopyridine-GO Diazopyridine-GO	Mild steel/1 M HCl	95.08/25 mg/L 96.73/25 mg/L	[105]
GO-sodium dodecyl sulfate GO-sodium propyl sulfate	Mild steel/1 M HCl	93.8%/200 mg/L 83.6%/200 mg/L	[106]
Diethylenetriamine-GO	Mild steel/1 M HCl	92.67%/25 mg/L	[107]
Bis (2-aminoethyl) amine-GO + KI	Carbon steel/15% HCl	96.77%/50 mg/L	[108]
Dopamine-GO	Carbon steel/15% HCl	80.69%/5 mg/L	[109]
Cyclodextrin-GO	X60 Carbon steel/15% HCl	68.61%/15% v/v	[113]
Polyethyleneimine-GO + KI	Carbon steel/15% HCl	95.77%/50 mg/L	[114]
Polyethyleneimine-GO	Copper/0.5 M HCl	92.24%/100 mg/L	[115]
GO-NSP	Mild steel/3.5% NaCl	100%/500 mg/L 100%/500 mg/L	[116]
Poly (o-anthranilic acid)/GO/ functionalized MWCNT	Stainless steel 316/2 M HCl	91.35%/70 mg/L	[117]
GON GOS	Carbon steel/1 M HCl	96.96%/6 mg/L 95.79%/6 mg/L	[118]
GO-quantum dots	Q235 steel/1 M HCl	88.38%/52.5 mg/L	[119]

7.5 Conclusions

This chapter presents a glimpse into the research work reported on Gr and Gr for anticorrosion coatings and corrosion inhibition. Single- and multilayered Gr has been extensively studied for their excellent barrier action. Gr coatings can effectively screen the underlying metallic substrate from the surrounding corrosive media. Although defects existing in the pure Gr film might accelerate the corrosion, another way to prepare the coatings is to use the chemically functionalized Gr or GO as fillers in the epoxy or polymer-based coatings. In such conditions, the GO can be covalently or noncovalently linked to the matrix components of the coatings. Several studies have been reported on the application of functionalized GO-based anticorrosion coatings. The coatings from Gr composites provide considerable corrosion resistance in comparison to conventional coatings due to their excellent barrier effect. In recent years, several studies have come across on the application of functionalized GO-based corrosion inhibitors for metals and alloys in aqueous acidic and saline environments. These studies have been reported in 0.5–2 M HCl and 15% HCl at room and elevated temperatures. These studies attempt to simulate the industrial acid-pickling and oil-well acidizing processes. Some studies are also available in 3.5% NaCl environments. In these cases, the functionalized GO-based inhibitors have shown excellent dispersibility in the aqueous media and have shown high inhibition efficiencies at low inhibitor concentrations. At prolonged immersion periods and at high temperatures, the modified GO-based corrosion inhibitors have been effective. Therefore, these inhibitors have come across as a fascinating alternative to the otherwise costly heterocyclic molecule-based corrosion inhibitors.

7.6 Prospects

In addition to Gr and GO, a number of other 2-D materials have been reported as anticorrosion coatings. Therefore, it is required to perform a thorough comparison of these materials to explore the potential application of Gr derivatives as corrosion protective agent. These studies can provide valuable insights on the replacement or possible fortification of these materials with Gr-based coatings. In the case of corrosion inhibitors based on Gr-based nanomaterials, more studies should be conducted on Gr, GO in the area of sweet corrosion. The detailed investigation should be conducted on the influence of temperature, pH, corrosion product formation, flow conditions and the influence of other additives. A prospective research area would be the study of GO-quantum dot (GOQD) systems. Some studies have come across in the literature on the application of carbon nanodots as corrosion inhibitors. Still, this is a potential area of further prospects.

References

[1] Fontana, M.G., Corrosion engineering, Tata McGraw-Hill Education, 2005.

[2] Revie, R.W., Uhlig's Corrosion Handbook, John Wiley & Sons, 2011.

[3] Quraishi, M.A., Chauhan, D.S., Saji, V.S., Heterocyclic Organic Corrosion Inhibitors: Principles and Applications, Elsevier Inc, Amsterdam, 2020, ISBN: 9780128185582.

[4] Sastri, V.S., Corrosion inhibitors: Principles and applications, John Wiley & Sons, 1998.

[5] Sastri, V.S., Green corrosion inhibitors: Theory and practice, John Wiley & Sons, 2012.

[6] Cizek, A., Corrosion inhibitors used in acidizing, Materials Performance;(United States), 1994, 33.

[7] Frenier, W., Growcock, F., Lopp, V., Mechanisms of corrosion inhibitors used in acidizing wells, SPE Production Engineering, 1988, 3, 584–590.

[8] Schmitt, G., Application of inhibitors for acid media: Report prepared for the European federation of corrosion working party on inhibitors, British Corrosion Journal, 1984, 19, 165–176.

[9] Finšgar, M., Jackson, J., Application of corrosion inhibitors for steels in acidic media for the oil and gas industry: A review, Corrosion Science, 2014, 86, 17–41.

[10] Finšgar, M., Milošev, I., Inhibition of copper corrosion by 1, 2, 3-benzotriazole: A review, Corrosion Science, 2010, 52, 2737–2749.

[11] Kokalj, A., Kovačević, N., Peljhan, S., Finšgar, M., Lesar, A., Milošev, I., Triazole, benzotriazole, and naphthotriazole as copper corrosion inhibitors: I, Molecular Electronic and Adsorption Properties, ChemPhysChem, 2011, 12, 3547–3555.

[12] Kokalj, A., Peljhan, S., Finsgar, M., Milosev, I., What determines the inhibition effectiveness of ATA, BTAH, and BTAOH corrosion inhibitors on copper?, Journal of the American Chemical Society, 2010, 132, 16657–16668.

[13] Xhanari, K., Finšgar, M., Organic corrosion inhibitors for aluminum and its alloys in chloride and alkaline solutions: A review, Arabian Journal of Chemistry, 2019, 12, 4646–4663.

[14] Sørensen, P.A., Kiil, S., Dam-Johansen, K., Weinell, C.E., Anticorrosive coatings: A review, Journal of Coatings Technology and Research, 2009, 6, 135–176.

[15] Verma, C., Quraishi, M.A., Chauhan, D.S., Green Corrosion Inhibition: Fundamentals, Design, Synthesis and Applications, Royal Society of Chemistry, 2022.

[16] Quraishi, M.A., Chauhan, D.S., Saji, V.S., Heterocyclic Organic Corrosion Inhibitors: Principles and Applications, Elsevier Inc. Amsterdam, 2020.

[17] Geim, A.K., Novoselov, K.S., The rise of graphene, in Nanoscience and Technology: A Collection of Reviews from Nature Journals, World Scientific, 2010, 11–19.

[18] Georgakilas, V., Perman, J.A., Tucek, J., Zboril, R., Broad family of carbon nanoallotropes: Classification, chemistry, and applications of fullerenes, carbon dots, nanotubes, graphene, nanodiamonds, and combined superstructures, Chemical Reviews, 2015, 115, 4744–4822.

[19] Allen, M.J., Tung, V.C., Kaner, R.B., Honeycomb carbon: A review of graphene, Chemical Reviews, 2009, 110, 132–145.

[20] Singh, V., Joung, D., Zhai, L., Das, S., Khondaker, S.I., Seal, S., Graphene based materials: Past, present and future, Progress in Materials Science, 2011, 56, 1178–1271.

[21] Ding, R., Li, W., Wang, X., Gui, T., Li, B., Han, P., Tian, H., Liu, A., Wang, X., Liu, X., A brief review of corrosion protective films and coatings based on graphene and graphene oxide, Journal of Alloys and Compounds, 2018, 764, 1039–1055.

[22] Kirkland, N., Schiller, T., Medhekar, N., Birbilis, N., Exploring graphene as a corrosion protection barrier, Corrosion Science, 2012, 56, 1–4.

[23] Wang, M., Tang, M., Chen, S., Ci, H., Wang, K., Shi, L., Lin, L., Ren, H., Shan, J., Gao, P., Graphene-Armored Aluminum Foil with Enhanced Anticorrosion Performance as Current Collectors for Lithium-Ion Battery, Advanced Materials, 2017, 29, 1703882.

[24] Gong, X., Liu, G., Li, Y., Yu, D.Y.W., Teoh, W.Y., Functionalized-graphene composites: Fabrication and applications in sustainable energy and environment, Chemistry of Materials, 2016, 28, 8082–8118.
[25] Georgakilas, V., Tiwari, J.N., Kemp, K.C., Perman, J.A., Bourlinos, A.B., Kim, K.S., Zboril, R., Noncovalent functionalization of graphene and graphene oxide for energy materials, biosensing, catalytic, and biomedical applications, Chemical Reviews, 2016, 116, 5464–5519.
[26] Cai, M., Thorpe, D., Adamson, D.H., Schniepp, H.C., Methods of graphite exfoliation, Journal of Materials Chemistry, 2012, 22, 24992–25002.
[27] Gogotsi, Y., Nanomaterials handbook, CRC press, 2006.
[28] Gao, W., The chemistry of graphene oxide, Graphene oxide, Springer, 2015, 61–95.
[29] Georgakilas, V., Otyepka, M., Bourlinos, A.B., Chandra, V., Kim, N., Kemp, K.C., Hobza, P., Zboril, R., Kim, K.S., Functionalization of graphene: Covalent and non-covalent approaches, derivatives and applications, Chemical Reviews, 2012, 112, 6156–6214.
[30] Saji, V.S., Cook, R., Corrosion protection and control using nanomaterials, Elsevier, 2012.
[31] Hu, J., Ji, Y., Shi, Y., Hui, F., Duan, H., Lanza, M., A review on the use of graphene as a protective coating against corrosion, Annals of Materials Science and Engineering, 2014, 1, 16.
[32] Cui, G., Bi, Z., Zhang, R., Liu, J., Yu, X., Li, Z., A comprehensive review on graphene-based anti-corrosive coatings, Chemical Engineering Journal, 2019, 373, 104–121.
[33] Iqbal, A.A., Sakib, N., Iqbal, A.P., Nuruzzaman, D.M., Graphene-based nanocomposites and their fabrication, mechanical properties and applications, Materialia, 2020, 12, 100815.
[34] Fishelson, N., Inberg, A., Croitoru, N., Shacham-Diamand, Y., Highly corrosion resistant bright silver metallization deposited from a neutral cyanide-free solution, Microelectronic Engineering, 2012, 92, 126–129.
[35] Pais, M., George, S.D., Rao, P., Glycogen nanoparticles as a potential corrosion inhibitor, International Journal of Biological Macromolecules, 2021, 182, 2117–2129.
[36] Sasikumar, Y., Kumar, A.M., Gasem, Z.M., Ebenso, E.E., Hybrid nanocomposite from aniline and CeO2 nanoparticles: Surface protective performance on mild steel in acidic environment, Applied Surface Science, 2015, 330, 207–215.
[37] Yoo, B.M., Shin, H.J., Yoon, H.W., Park, H.B., Graphene and graphene oxide and their uses in barrier polymers, Journal of Applied Polymer Science, 2014, 131.
[38] Li, J., Cui, J., Yang, J., Ma, Y., Qiu, H., Yang, J., Silanized graphene oxide reinforced organofunctional silane composite coatings for corrosion protection, Progress in Organic Coatings, 2016, 99, 443–451.
[39] Tong, Y., Bohm, S., Song, M., Graphene based materials and their composites as coatings, Austin Journal of Nanomedicine & Nanotechnology, 2013, 1, 1003.
[40] Li, X., Cai, W., An, J., Kim, S., Nah, J., Yang, D., Piner, R., Velamakanni, A., Jung, I., Tutuc, E., Large-area synthesis of high-quality and uniform graphene films on copper foils, Science, 2009, 324, 1312–1314.
[41] Nan, L., Wang, Z., Zhao, K., Shi, Z., Gu, Z., Xu, S., Large scale synthesis of N-doped multi-layered graphene sheets by simple arc-discharge method, Carbon, 2010, 48, 255–259.
[42] Karmakar, S., Kulkarni, N.V., Nawale, A.B., Lalla, N.P., Mishra, R., Sathe, V.G., Bhoraskar, S.V., Das, A.K., A novel approach towards selective bulk synthesis of few-layer graphenes in an electric arc, Journal of Physics D: Applied Physics, 2009, 42, 1–21.
[43] Chen, S., Brown, L., Levendorf, M., Cai, W., Ju, S.-Y., Edgeworth, J., Li, X., Magnuson, C.W., Velamakanni, A., Piner, R.D., Oxidation resistance of graphene-coated Cu and Cu/Ni alloy, ACS Nano, 2011, 5, 1321–1327.
[44] Yu, L., Lim, Y.-S., Han, J.H., Kim, K., Kim, J.Y., Choi, S.-Y., Shin, K., A graphene oxide oxygen barrier film deposited via a self-assembly coating method, Synthetic Metals, 2012, 162, 710–714.
[45] Weatherup, R.S., D'Arsié, L., Cabrero-Vilatela, A., Caneva, S., Blume, R., Robertson, J., Schloegl, R., Hofmann, S., Long-term passivation of strongly interacting metals with single-layer graphene, Journal of the American Chemical Society, 2015, 137, 14358–14366.

[46] Hsieh, Y.-P., Hofmann, M., Chang, K.-W., Jhu, J.G., Li, -Y.-Y., Chen, K.Y., Yang, C.C., Chang, W.-S., Chen, L.-C., Complete corrosion inhibition through graphene defect passivation, ACS Nano, 2013, 8, 443–448.

[47] Guermoune, A., Chari, T., Popescu, F., Sabri, S.S., Guillemette, J., Skulason, H.S., Szkopek, T., Siaj, M., Chemical vapor deposition synthesis of graphene on copper with methanol, ethanol, and propanol precursors, Carbon, 2011, 49, 4204–4210.

[48] Raman, R.S., Banerjee, P.C., Lobo, D.E., Gullapalli, H., Sumandasa, M., Kumar, A., Choudhary, L., Tkacz, R., Ajayan, P.M., Majumder, M., Protecting copper from electrochemical degradation by graphene coating, Carbon, 2012, 50, 4040–4045.

[49] Prasai, D., Tuberquia, J.C., Harl, R.R., Jennings, G.K., Bolotin, K.I., Graphene: Corrosion-inhibiting coating, ACS Nano, 2012, 6, 1102–1108.

[50] Huh, J.-H., Kim, S.H., Chu, J.H., Kim, S.Y., Kim, J.H., Kwon, S.-Y., Enhancement of seawater corrosion resistance in copper using acetone-derived graphene coating, Nanoscale, 2014, 6, 4379–4386.

[51] Lee, J., Berman, D., Inhibitor or promoter: Insights on the corrosion evolution in a graphene protected surface, Carbon, 2018, 126, 225–231.

[52] Anisur, M., Banerjee, P.C., Easton, C.D., Raman, R.S., Controlling hydrogen environment and cooling during CVD graphene growth on nickel for improved corrosion resistance, Carbon, 2018, 127, 131–140.

[53] Halkjær, S., Iversen, J., Kyhl, L., Chevallier, J., Andreatta, F., Yu, F., Stoot, A., Camilli, L., Bøggild, P., Hornekær, L., Low-temperature synthesis of a graphene-based, corrosion-inhibiting coating on an industrial grade alloy, Corrosion Science, 2019, 152, 1–9.

[54] Mišković-Stanković, V., Jevremović, I., Jung, I., Rhee, K., Electrochemical study of corrosion behavior of graphene coatings on copper and aluminum in a chloride solution, Carbon, 2014, 75, 335–344.

[55] Ye, X., Lin, Z., Zhang, H., Zhu, H., Liu, Z., Zhong, M., Protecting carbon steel from corrosion by laser in situ grown graphene films, Carbon, 2015, 94, 326–334.

[56] Pu, N.-W., Shi, G.-N., Liu, Y.-M., Sun, X., Chang, J.-K., Sun, C.-L., Ger, M.-D., Chen, C.-Y., Wang, P.-C., Peng, -Y.-Y., Graphene grown on stainless steel as a high-performance and ecofriendly anti-corrosion coating for polymer electrolyte membrane fuel cell bipolar plates, Journal of Power Sources, 2015, 282, 248–256.

[57] Stoot, A.C., Camilli, L., Spiegelhauer, S.-A., Yu, F., Bøggild, P., Multilayer graphene for long-term corrosion protection of stainless steel bipolar plates for polymer electrolyte membrane fuel cell, Journal of Power Sources, 2015, 293, 846–851.

[58] Ding, R., Zheng, Y., Yu, H., Li, W., Wang, X., Gui, T., Study of water permeation dynamics and anti-corrosion mechanism of graphene/zinc coatings, Journal of Alloys and Compounds, 2018, 748, 481–495.

[59] Zhang, L., Duan, Y., Gao, Z., Ma, J., Liu, R., Liu, S., Tu, Z., Liu, Y., Bai, C., Cui, L., Graphene enhanced anti-corrosion and biocompatibility of NiTi alloy, NanoImpact, 2017, 7, 7–14.

[60] Noël, S., Baraton, L., Alamarguy, D., Jaffré, A., Viel, P., Palacin, S., Hauquier, F., Graphene films for corrosion protection of gold coated cuprous substrates in view of an application to electrical contacts, in: 2012 IEEE 58th Holm Conference on Electrical Contacts (Holm), IEEE, 2012, 1–7.

[61] Yivlialin, R., Bussetti, G., Duò, L., Yu, F., Galbiati, M., Camilli, L., CVD Graphene/Ni interface evolution in sulfuric electrolyte, Langmuir, 2018, 34, 3413–3419.

[62] Zhu, M., Du, Z., Yin, Z., Zhou, W., Liu, Z., Tsang, S.H., Teo, E.H.T., Low-temperature in situ growth of graphene on metallic substrates and its application in anticorrosion, ACS Applied Materials & Interfaces, 2015, 8, 502–510.

[63] Ye, X., Yu, F., Curioni, M., Lin, Z., Zhang, H., Zhu, H., Liu, Z., Zhong, M., Corrosion resistance of graphene directly and locally grown on bulk nickel substrate by laser irradiation, RSC Advances, 2015, 5, 35384–35390.

[64] Lanza, M., Wang, Y., Sun, H., Tong, Y., Duan, H., Morphology and performance of graphene layers on as-grown and transferred substrates, Acta Mechanica, 2014, 225, 1061–1073.

[65] Tiwari, A., Singh Raman, R., Durable corrosion resistance of copper due to multi-layer graphene, Materials, 2017, 10, 1112.

[66] Kang, D., Kwon, J.Y., Cho, H., Sim, J.-H., Hwang, H.S., Kim, C.S., Kim, Y.J., Ruoff, R.S., Shin, H.S., Oxidation resistance of iron and copper foils coated with reduced graphene oxide multilayers, Acs Nano, 2012, 6, 7763–7769.

[67] Chauhan, D.S., Quraishi, M., Ansari, K., Saleh, T.A., Graphene and graphene oxide as new class of materials for corrosion control and protection: Present status and future scenario, Progress in Organic Coatings, 2020, 147, 105741.

[68] Dong, Y., Liu, Q., Zhou, Q., Corrosion behavior of Cu during graphene growth by CVD, Corrosion Science, 2014, 89, 214–219.

[69] Yu, F., Camilli, L., Wang, T., Mackenzie, D.M., Curioni, M., Akid, R., Bøggild, P., Complete long-term corrosion protection with chemical vapor deposited graphene, Carbon, 2018, 132, 78–84.

[70] Mohammadi, S., Roohi, H., Influence of functionalized multi-layer graphene on adhesion improvement and corrosion resistance performance of zinc-rich epoxy primer, Corrosion Engineering, Science and Technology, 2018, 53, 422–430.

[71] Compton, O.C., Kim, S., Pierre, C., Torkelson, J.M., Nguyen, S.T., Crumpled graphene nanosheets as highly effective barrier property enhancers, Advanced Materials, 2010, 22, 4759–4763.

[72] Vadukumpully, S., Paul, J., Mahanta, N., Valiyaveettil, S., Flexible conductive graphene/poly (vinyl chloride) composite thin films with high mechanical strength and thermal stability, Carbon, 2011, 49, 198–205.

[73] Chen, C., Qiu, S., Cui, M., Qin, S., Yan, G., Zhao, H., Wang, L., Xue, Q., Achieving high performance corrosion and wear resistant epoxy coatings via incorporation of noncovalent functionalized graphene, Carbon, 2017, 114, 356–366.

[74] Kim, H., Miura, Y., Macosko, C.W., Graphene/polyurethane nanocomposites for improved gas barrier and electrical conductivity, Chemistry of Materials, 2010, 22, 3441–3450.

[75] Kim, H., Abdala, A.A., Macosko, C.W., Graphene/polymer nanocomposites, Macromolecules, 2010, 43, 6515–6530.

[76] Kim, H., Macosko, C.W., Processing-property relationships of polycarbonate/graphene composites, Polymer, 2009, 50, 3797–3809.

[77] Kim, H., Macosko, C.W., Morphology and properties of polyester/exfoliated graphite nanocomposites, Macromolecules, 2008, 41, 3317–3327.

[78] Potts, J.R., Dreyer, D.R., Bielawski, C.W., Ruoff, R.S., Graphene-based polymer nanocomposites, Polymer, 2011, 52, 5–25.

[79] Sun, W., Wang, L., Wu, T., Wang, M., Yang, Z., Pan, Y., Liu, G., Inhibiting the corrosion-promotion activity of graphene, Chemistry of Materials, 2015, 27, 2367–2373.

[80] Liang, J., Wu, X.-W., Ling, Y., Yu, S., Zhang, Z., Trilaminar structure hydrophobic graphene oxide decorated organosilane composite coatings for corrosion protection, Surface and Coatings Technology, 2018, 339, 65–77.

[81] Mo, M., Zhao, W., Chen, Z., Liu, E., Xue, Q., Corrosion inhibition of functional graphene reinforced polyurethane nanocomposite coatings with regular textures, Rsc Advances, 2016, 6, 7780–7790.

[82] Hayatdavoudi, H., Rahsepar, M., A mechanistic study of the enhanced cathodic protection performance of graphene-reinforced zinc rich nanocomposite coating for corrosion protection of carbon steel substrate, Journal of Alloys and Compounds, 2017, 727, 1148–1156.

[83] Gu, L., Liu, S., Zhao, H., Yu, H., Facile preparation of water-dispersible graphene sheets stabilized by carboxylated oligoanilines and their anticorrosion coatings, ACS Applied Materials & Interfaces, 2015, 7, 17641–17648.

[84] Monetta, T., Acquesta, A., Bellucci, F., Graphene/epoxy coating as multifunctional material for aircraft structures, Aerospace, 2015, 2, 423–434.

[85] Li, X., Bandyopadhyay, P., Guo, M., Kim, N.H., Lee, J.H., Enhanced gas barrier and anticorrosion performance of boric acid induced cross-linked poly (vinyl alcohol-co-ethylene)/graphene oxide film, Carbon, 2018, 133, 150–161.

[86] Parhizkar, N., Ramezanzadeh, B., Shahrabi, T., Corrosion protection and adhesion properties of the epoxy coating applied on the steel substrate pre-treated by a sol-gel based silane coating filled with amino and isocyanate silane functionalized graphene oxide nanosheets, Applied Surface Science, 2018, 439, 45–59.

[87] Yu, Z., Di, H., Ma, Y., Lv, L., Pan, Y., Zhang, C., He, Y., Fabrication of graphene oxide–alumina hybrids to reinforce the anti-corrosion performance of composite epoxy coatings, Applied Surface Science, 2015, 351, 986–996.

[88] Chang, K.-C., Ji, W.-F., Lai, M.-C., Hsiao, Y.-R., Hsu, C.-H., Chuang, T.-L., Wei, Y., Yeh, J.-M., Liu, W.-R., Synergistic effects of hydrophobicity and gas barrier properties on the anticorrosion property of PMMA nanocomposite coatings embedded with graphene nanosheets, Polymer Chemistry, 2013, 5, 1049–1056.

[89] Aneja, K.S., Bohm, S., Khanna, A., Bohm, H.M., Graphene based anticorrosive coatings for Cr (VI) replacement, Nanoscale, 2015, 7, 17879–17888.

[90] Zhan, Y., Zhang, J., Wan, X., Long, Z., He, S., He, Y., Epoxy composites coating with Fe_3O_4 decorated graphene oxide: Modified bio-inspired surface chemistry, synergistic effect and improved anti-corrosion performance, Applied Surface Science, 2018, 436, 756–767.

[91] Kumar, A., Anant, R., Kumar, K., Chauhan, S.S., Kumar, S., Kumar, R., Anticorrosive and electromagnetic shielding response of a graphene/TiO_2–epoxy nanocomposite with enhanced mechanical properties, RSC Advances, 2016, 6, 113405–113414.

[92] Di, H., Yu, Z., Ma, Y., Pan, Y., Shi, H., Lv, L., Li, F., Wang, C., Long, T., He, Y., Anchoring calcium carbonate on graphene oxide reinforced with anticorrosive properties of composite epoxy coatings, Polymers for Advanced Technologies, 2016, 27, 915–921.

[93] Yang, S.-Y., Lin, W.-N., Huang, Y.-L., Tien, H.-W., Wang, J.-Y., Ma, -C.C.M., Li, S.-M., Wang, Y.-S., Synergetic effects of graphene platelets and carbon nanotubes on the mechanical and thermal properties of epoxy composites, Carbon, 2011, 49, 793–803.

[94] Liu, J., Yu, Q., Yu, M., Li, S., Zhao, K., Xue, B., Zu, H., Silane modification of titanium dioxide-decorated graphene oxide nanocomposite for enhancing anticorrosion performance of epoxy coatings on AA-2024, Journal of Alloys and Compounds, 2018, 744, 728–739.

[95] Zhou, S., Wu, Y., Zhao, W., Yu, J., Jiang, F., Wu, Y., Ma, L., Designing reduced graphene oxide/zinc rich epoxy composite coatings for improving the anticorrosion performance of carbon steel substrate, Materials & Design, 2019, 169, 107694.

[96] Zhang, L., Wu, H., Zheng, Z., He, H., Wei, M., Huang, X., Fabrication of graphene oxide/multi-walled carbon nanotube/urushiol formaldehyde polymer composite coatings and evaluation of their physico-mechanical properties and corrosion resistance, Progress in Organic Coatings, 2019, 127, 131–139.

[97] Hu, H., He, Y., Long, Z., Zhan, Y., Synergistic effect of functional carbon nanotubes and graphene oxide on the anti-corrosion performance of epoxy coating, Polymers for Advanced Technologies, 2017, 28, 754–762.

[98] Sun, W., Wang, L., Wu, T., Pan, Y., Liu, G., Inhibited corrosion-promotion activity of graphene encapsulated in nanosized silicon oxide, Journal of Materials Chemistry A, 2015, 3, 16843–16848.

[99] Yu, Y.-H., Lin, -Y.-Y., Lin, C.-H., Chan, -C.-C., Huang, Y.-C., High-performance polystyrene/graphene-based nanocomposites with excellent anti-corrosion properties, Polymer Chemistry, 2014, 5, 535–550.

[100] Li, X., Bandyopadhyay, P., Nguyen, T.T., Park, O.-K., Lee, J.H., Fabrication of functionalized graphene oxide/maleic anhydride grafted polypropylene composite film with excellent gas barrier and anticorrosion properties, Journal of Membrane Science, 2018, 547, 80–92.

[101] Yang, Z., Wang, L., Sun, W., Li, S., Zhu, T., Liu, W., Liu, G., Superhydrophobic epoxy coating modified by fluorographene used for anti-corrosion and self-cleaning, Applied Surface Science, 2017, 401, 146–155.

[102] Chun Kiang Chua, M.P., Covalent chemistry on graphene, Chemical Society Reviews, 2013, 42, 3222–3233.

[103] Gupta, R.K., Malviya, M., Verma, C., Quraishi, M.A., Aminoazobenzene and diaminoazobenzene functionalized graphene oxides as novel class of corrosion inhibitors for mild steel: Experimental and DFT studies, Materials Chemistry and Physics, 2017, 198, 360–373.

[104] Gupta, R.K., Malviya, M., Ansari, K.R., Lgaz, H., Chauhan, D.S., Quraishi, M.A., Functionalized graphene oxide as a new generation corrosion inhibitor for industrial pickling process: DFT and experimental approach, Materials Chemistry and Physics, 2019, 236, 121727.

[105] Gupta, R.K., Malviya, M., Verma, C., Gupta, N.K., Quraishi, M.A., Pyridine-based functionalized graphene oxides as a new class of corrosion inhibitors for mild steel: An experimental and DFT approach, RSC Advances, 2017, 7, 39063–39074.

[106] Ansari, K., Chauhan, D.S., Quraishi, M., Saleh, T.A., Surfactant modified graphene oxide as novel corrosion inhibitors for mild steels in acidic media, Inorganic Chemistry Communications, 2020, 121, 108238.

[107] Baig, N., Chauhan, D.S., Saleh, T.A., Quraishi, M.A., Diethylenetriamine functionalized graphene oxide as a novel corrosion inhibitor for mild steel in hydrochloric acid solutions, New Journal of Chemistry, 2019, 43, 2328–2337.

[108] Ansari, K.R., Chauhan, D.S., Quraishi, M.A., Saleh, T.A., Bis (2-aminoethyl) amine-modified graphene oxide nanoemulsion for carbon steel protection in 15% HCl: Effect of temperature and synergism with iodide ions, Journal of Colloid and Interface Science, 2020, 564, 124–133.

[109] Haruna, K., Alhems, L.M., Saleh, T.A., Graphene oxide grafted with dopamine as an efficient corrosion inhibitor for oil well acidizing environments, Surfaces and Interfaces, 2021, 24, 101046.

[110] Goni, L.K., Mazumder, M.A.J., Ali, S.A., Chauhan, D.S., Water-soluble polymeric corrosion inhibitors, in Mazumder, M.J., Quraishi, M., Al-Ahmed, A. (Eds.), polymeric corrosion inhibitors for greening the chemical and petrochemical industry, Wiley-VCH GmbH, 2023, 97–123.

[111] Chauhan, D.S., Srivastava, V., Lin, Y., Quraishi, M.A., Polymers as Corrosion Inhibitors for Sweet Environment, in Mazumder, M.J., Quraishi, M., Al-Ahmed, A. (Eds.), polymeric corrosion inhibitors for greening the chemical and petrochemical industry, Wiley-VCH GmbH, 2023, 193–220.

[112] Chauhan, D.S., Quraishi, M.A., Al-Qahtani, H., Mazumder, M.A.J., Green Polymeric Corrosion Inhibitors: Design, Synthesis, and Characterization, in Mazumder, M.J., Quraishi, M., Al-Ahmed, A. (Eds.), Polymeric corrosion inhibitors for greening the chemical petrochemical industry, Wiley-VCH GmbH, 2023, 1–22.

[113] Haruna, K., Saleh, T.A., Obot, I., Umoren, S.A., Cyclodextrin-based functionalized graphene oxide as an effective corrosion inhibitor for carbon steel in acidic environment, Progress in Organic Coatings, 2019, 128, 157–167.

[114] Ansari, K., Chauhan, D.S., Quraishi, M.A., Adesina, A., Saleh, T.A., The synergistic influence of polyethyleneimine-grafted graphene oxide and iodide for the protection of steel in acidizing conditions, RSC Advances, 2020, 10, 17739–17751.

[115] Quraishi, M.A., Singh Chauhan, D., Rahmani Ansari, K., Madhan Kumar, A., Saleh, T.A., Polyethyleneimine functionalized graphene oxide: a promising inhibitor for corrosion of copper in the hydrochloric acid environment, Journal of Nanoscience and Nanotechnology, 2021, 21, 3256–3268.

[116] Sharifi, Z., Pakshir, M., Amini, A., Rafiei, R., Hybrid graphene oxide decoration and water-based polymers for mild steel surface protection in saline environment, Journal of Industrial and Engineering Chemistry, 2019, 74, 41–54.

[117] Shirazi, Z., Golikand, A.N., Keshavarz, M.H., A new nanocomposite based on poly (o-anthranilic acid), graphene oxide and functionalized carbon nanotube as an efficient corrosion inhibitor for stainless steel in severe environmental corrosion, Composites Communications, 2020, 22, 100467.

[118] Radey, H.H., Khalaf, M.N., Al-Sawaad, H.Z., novel corrosion inhibitors for carbon steel alloy in acidic medium of 1N HCl synthesized from graphene oxide, Open Journal of Organic Polymer Materials, 2018, 8, 53.

[119] Chen, Z., Wang, M., Fadhil, A.A., Fu, C., Chen, T., Chen, M., Khadom, A.A., Mahood, H.B., Preparation, characterization, and corrosion inhibition performance of graphene oxide quantum dots for Q235 steel in 1 M hydrochloric acid solution, Colloids and Surfaces A, Physicochemical and Engineering Aspects, 2021, 627, 127209.

Dakeshwar kumar Verma*, Reema Sahu, Santosh Bahadur Singh*,
Bharti Yarda, Vikas Kumar Jain, Shailendra Yadav, Vikash Kande,
Sharad Tiwari, Gokul Ram Nishad, Younus Raza Beg, Vandana Mishra,
Durgesh Sinha

Chapter 8
Carbon dots (CDs) and heteroatom-doped CDs in corrosion prevention

Abstract: With diameters about 10–100 nm, carbon dots (CDs) and heteroatom-doped CDs are intriguing types of carbon nanoparticles. Low toxicity, photo-induced electron transfer, chemical inertness, good biocompatibility and highly controllable photoluminescence behavior are some special qualities of CDs and heteroatom-doped CDs. Due to their affordability, environmental friendliness and ability to reduce waste generation, sustainable raw materials are frequently employed in the production of CDs and heteroatom-doped CDs. Laser ablation, electrochemical oxidation, hydrothermal reaction, microwave irradiation, reflux technique and ultrasonication can all be used to synthesize CDs and heteroatom-doped CDs. Additionally, it has highly desirable characteristics like semiconductor nanoparticles and oxygen-based functional groups. As a result, CDs are promising nanomaterials for applications such as photo-catalysis, ion sensing, biological imaging, heavy metal detection, adsorption treatment, supercapacitor, membrane construction and water pollution control. The physical and chemical characteristics of CDs, the raw materials and processes employed in their production, their stability and their

Acknowledgments: The authors thank Dr. K. L. Tandekar, Principal, Govt Digvijay Autonomous PG College, Rajnandgaon, Chhattisgarh, for providing basic facilities.

*Corresponding author: Dakeshwar kumar Verma**, Department of Chemistry, Government Digvijay Autonomous Postgraduate College, Rajnandgaon 491441, Chhattisgarh, India,
e-mail: dakeshwarverma@gmail.com
*Corresponding author: Santosh Bahadur Singh**, Department of Chemistry, University of Allahabad, Prayagraj 211002, Uttar Pradesh, India, e-mail: singhsbau2012@gmail.com
Reema Sahu, Bharti Yarda, Vikash Kande, Sharad Tiwari, Gokul Ram Nishad, Younus Raza Beg, Vandana Mishra, Department of Chemistry, Government Digvijay Autonomous Postgraduate College, Rajnandgaon 491441, Chhattisgarh, India
Vikas Kumar Jain, Department of Technical Education Indravati Bhawan, Nava Raipur, Atal Nagar, Raipur 492002, Chhattisgarh, India
Shailendra Yadav, Department of Chemistry, AKS University, Satna, Madhya Pradesh, India
Durgesh Sinha, Department of Chemistry, Gurughasidas Central University, Bilaspur, Chhattisgarh, India

https://doi.org/10.1515/9783111071756-008

prospective uses in the prevention of corrosion of metals in various acidic conditions will all be covered in this review.

Keywords: Carbon dots, corrosion inhibition, mild steel, acidic solution, electrochemical study, SEM-EDS, DFT

8.1 Introduction

Corrosion-induced metal deterioration can lead to significant financial loss and compromised life safety. Acidification of oil wells, chemical cleaning and processing, pickling and so on include the use of common acid-aggressive media including sulfuric acid, hydrochloric acid, phosphoric acid and hydrofluoric acid [1, 2]. The metal substrate's protective approach is crucial during the aforementioned operations. The use of corrosion inhibitors is one of many anticorrosion techniques that is thought to be a successful, easy-to-use and inexpensive tactic for delaying or reducing the corrosion of the steel substrate. The structural properties of an inhibitor, the type of metal surface, the harsh environment and so on affect how effective it is at inhibition [3]. Some conventional high-efficiency corrosion inhibitors, like chromate and compounds containing phosphorus, are hazardous to some extent and have a difficult time degrading in the environment [4]. In the accelerated growth of industrial technology, which affects numerous national production domains, metal corrosion has been a global issue [5]. According to study corrosion costs the world economy up to 3% of GDP annually and results in the loss of roughly one-third of the steel produced [6]. As a result, the industry has developed a variety of methods for protecting against metal corrosion. One of the simplest ways to prevent metals from corroding in corrosive environments is to use a corrosion inhibitor. As a result, the use of organic corrosion inhibitors is becoming increasingly important in industrial areas. As effective corrosion inhibitors for metals in aggressive environments, organic compounds having N, O, P and S as well as an aromatic ring and a long alkyl chain in their structural component have been reported [7, 8]. These corrosion inhibitors provide effective inhibition, but some of them are toxic and bad for the environment and human health. The demand for environmentally friendly pickling corrosion inhibitors with high corrosion inhibition efficiency and low price has increased as a result of people's heightened awareness of the environment [9, 10]. Nanosized semiconductor crystals, such as carbon dots (CDs) and heteroatom-doped CDs, are typically organic in origin. Due to the fact that poisonous heavy metals are primarily used in their production, inorganic quantum dots are subject to numerous safety and toxicity problems. Due to the fact that they are both organic and inorganic in nature and are widely employed as anticorrosion materials, the recently produced CDs and heteroatom-doped CDs can solve this drawback. Additionally Table 8.1 reveals the global GDP decline caused by the corrosion and Table 8.2 shows the development of corrosion inhibitors for aqueous electrolytes.

Table 8.1: The global GDP decline brought caused by rust (source: http://impact.nace.org/economic-im pact.aspx).

Monetary region	Agri CoC ($ billion)	Industry CoC ($ billion)	Services CoC ($ billion)	Total CoC ($ billion)	Total GDP ($ billion)	CoC % GDP
Arab World	13.3	34.2	92.6	140.1	2,789	5.0%
China	56.2	192.5	146.2	394.9	9,330	4.2%
European region	3.5	401	297	701.5	18,331	3.8%
India	17.7	20.3	32.3	70.3	1,670	4.2%
Japan	0.6	45.9	5.1	51.6	5,002	1.0%
Russia	5.4	37.2	41.9	84.5	2,113	4.0%
USA	2.0	303.2	146.0	451.3	16,720	2.7%
Other part of the globe	52.4	382.5	117.6	552.5	16,057	3.4%
Global	**152.7**	**1,446.7**	**906.0**	**2,505.4**	**74,314**	**3.4%**

Table 8.2: Corrosion inhibitors for aqueous electrolytes are being developed [11].

Time edge	Need	Examples (inhibitor type)
Before 1960	Protection efficiency	Borates, silicates, phosphates, borates, chromates, zinc salts and phosphates
Between 1960 and 1980	Economy	Phosphono acids, molybdates, polyphosphates, chelate molecules, carboxylates, gluconates, soluble oils, cations and polyacrylates are some examples of chemical compounds
Between 1980 and 1995	Ecology	Inhibitors found in nature including vitamins, biopolymers and tannins
From 1995 to present	Green and sustainability	REM, synergistic organic/inorganic compounds, inhibitor encapsulation and polyfunctional organic molecules

8.2 CDs and heteroatom-doped CDs: preparative methods, properties, and recent applications

The pyrolysis approach was used to produce citric acid-based CDs in the study at various reaction temperatures. The major goal was to investigate the connections between reaction temperature, microstructure and the capacity of as-obtained N-CDs to suppress corrosion in a 1 M HCl environment. Meanwhile, the corrosion inhibition mechanism of the as-obtained N-CDs was thoroughly analyzed using molecular dynamics and quantum chemical simulation calculations. In the recent years, we have seen a rise in interest in CDs (CDs), a type of carbon-based fluorescent (FL) nanomaterial with minimal cytotoxicity, high water solubility and exceptional corrosion inhibition capabilities. Different natural carbon sources including orange juice [12], citric acid [13], bananas [14] and red chillies [15] have all been used to create CDs. Hydrothermal carbonization, arc discharge, electrochemical synthesis, microwave approach and laser ablation have all been used to prepare CDs [16–18]. Inorganic nutrients and hazardous metals may taint CDs made from natural sources. Spice-derived CDs do not display cytotoxicity, making them suitable for application in the development of cancer therapy protocols or drug delivery management techniques [15]. Researchers study on the behavior and mechanism of N-doped CDs for metal in acid environments. These N-CDs' inhibitive efficacy in this instance was greater than 90%. Calculations revealed that the as-prepared N-CDs' standard adsorption free energies were −25.99, −26.94 and −25.78 kJ/mol, indicating that both chemisorption and physisorption were engaged in the adsorption film [19]. In addition, Figure 8.1 exhibits the synthesis method of NCDs.

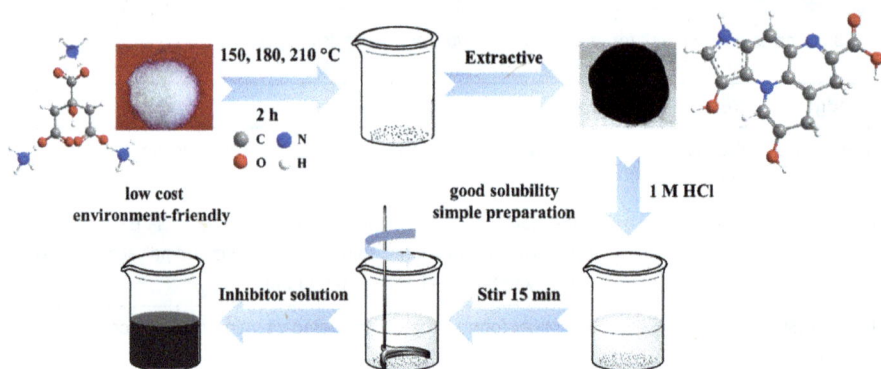

Figure 8.1: Synthesis methods of NCDs [19].

Before being placed into polytetrafluoroethylene autoclaves and heated to high temperature for long time, 4-aminosalicylic acid (ASA) was first dissolved in known amount of ethanol while being stirred. After cooling to ambient temperature, dark brown solutions were obtained. The obtained solution was purified for one day using dialysis bags to get rid of nonreactive molecules. After rotary evaporation and drying under vacuum, N-CDs were eventually obtained as a black solid and have greater solubility in ethanol. DI water should be replaced every 3 h [20].

8.3 CDs and heteroatom-doped CDs as advanced anticorrosive materials: experimental and computational approaches

It is discovered that CDs, a novel family of nanomaterials, are effective inhibitors. Given that there is no precise mechanical explanation in the literature for how CDs prevent carbon steel from corroding in an aggresive environment. Eco-friendly CDs and nitrogen-doped functionalized carbon dots (N-CDs) were easily made by reacting citric acid and urea, two inexpensive raw materials with several functional groups, in a single hydrothermal step. Electrochemical techniques and surface analysis were used to carefully explore their corrosion inhibition characteristics. For carbon steel in an acidic solution, the resulting CDs and N-CDs were found to be an efficient corrosion inhibitor. The Langmuir adsorption model and the N-doped CDs' 94% inhibitory efficacy at 200 mg/L show that the N-doped CDs under investigation are mixed type [21]. At 200 mg/L, N-CDs had the highest inhibition efficiency (96.13%), followed by the Langmuir model [22]. At 200 mg/L, CDs had an IE of over 90.9%, the inhibitor displayed the Langmuir adsorption model and the steel/solution interface underwent both chemical and physical adsorptions [23]. The %IE for p-CDs and o-CDs was more than 97% and changed gradually as immersion duration increased. The absorption of the p-CDs and o-CDs involved both chemisorption and physisorption according to the Langmuir adsorption isotherm [24]. At N-CDs 298 K at 30 mg/L, a percentage inhibition efficiency of 97.8% was obtained. Various adsorption models were looked at, and it was discovered that both physical and chemical adsorption took place at the steel/solution interface [25]. In a similar approach the % IE of N, S-CDs achieved 85.9% at 5 mg/L [26]. N, S-CDs is a mixed-type inhibitor, more obviously blocking the anodic reaction, with % IE of 93% at 50 mg/L [27]. The Langmuir adsorption model, in which physical contact was predominate, was used, and the % IE of N-doped CDs was 88.96% at 200 mg/L [28]. The physicochemical interaction at the steel/solution interface was exploited as the adsorption mechanism, and the % IE of FCDs was above 90% at 100 mg/L [29]. Additionally, Table 8.3 demonstrates the CDs-mediated corrosion inhibitor, electrolyte and metal sample, techniques applied and outcomes of CDs and doped materials as anticorrosion materials.

Despite the addition of various quantities of N, S-CDs, the capacitive loops grew during the first 3 h and then gradually shrank as immersion duration increased as seen in Figure 8.3(c–f). Additionally, the difference between the impedance spectra measured at various immersion periods is inversely proportional to the inhibitor concentrations, with the least change occurring at 5 mg/L N, S-CDs. The numerous inductive loops are responsible for the Bode plots' low-frequency display of various time constants in the blank condition (Figure 8.2) [26].

Via ASA as a precursor, the N-CDs were produced using a solvothermal process. According to scanning probe microscopy and transmission electron microscopy, N-CDs

Table 8.3: CD-mediated corrosion inhibitor, electrolyte and metal sample, techniques applied, and outcomes of CDs and doped materials as anticorrosion materials.

CDs-mediated corrosion inhibitor	Electrolyte and metal sample	Techniques applied	Outcomes	Reference
N-doped carbon dots	1 M HCl/steel	FTIR, UV–vis, XPS, EIS and WL	The Langmuir adsorption model and the N-doped carbon dots' (CDs') 94% inhibitory efficacy at 200 mg/L show that the N-doped CDs under investigation are mixed type	[21]
N-doped carbon dots (NCDs)	0.1 M HCl/Q235 steel	FTIR, UV–vis, XRF, XPS, OCP, Tafel, EIS analysis and MD simulation	At 200 mg/L, N-CDs had the highest inhibition efficiency (96.13%), followed by the Langmuir model	[22]
N-doped carbon dots (CDs)	1 M HCl/Q235 steel	FTIR, UV–vis, XPS, TEM, SPM, OCP, EIS, WL and zeta potential analysis	At 200 mg/L, CDs had an IE of over 90.9%, the inhibitor displayed the Langmuir adsorption model, and the steel/solution interface underwent both chemical and physical adsorptions	[23]
p-CDs and o-CDs	1 M HCl Q235/ carbon steel	PDp, EIS, SEM and EDX	The % IE for p-CDs and o-CDs was more than 97% and changed gradually as immersion duration increased. The absorption of the p-CDs and o-CDs involved both chemisorption and physisorption according to the Langmuir adsorption isotherm.	[24]
Nitrogen-doped functionalized carbon dots (NCDs)	0.5 M H$_2$SO$_4$/ carbon steel	FTIR, UV–vis, XRD, TEM, EIS, PDP, AFM, SEM and XPS	At N-CDs 298 K at 30 mg/L, a percentage inhibition efficiency of 97.8% was obtained. Various adsorption models were looked at, and it was discovered that both physical and chemical adsorption took place at the steel/solution interface	[25]
N and S codoped carbon dots (N, SCDs)	0.1 M HCl/5,052 Al alloy	XPS, WL, OCP, PDP, SEM, IFM, FTIR, AFM, weight	% IE of N, S-CDs achieved 85.9% at 5 mg/L	[26]

N and S-codoped carbon dots (N, S-CDs)	CO_2-saturated 3% NaCl/carbon steel	WL, XPS, SEM, AFM, contact angle, OCP, PDP and EIS	N, S-CDs is a mixed-type inhibitor, more obviously blocking the anodic reaction, with % IE of 93% at 50 mg/L	[27]
N-doped carbon dots	3.5% NaCl/ carbon steel	EIS, WL, and corrosion morphology	The Langmuir adsorption model, in which physical contact was predominate, was used, and the % IE of N-doped carbon dots was 88.96% at 200 mg/L	[28]
Functionalized carbon dots (FCDs)	1 M HCl Q235/ steel	OCP, EIS, Tafel, SVET and corrosion morphology analysis	The physicochemical interaction at the steel/solution interface was exploited as the adsorption mechanism, and the % IE of FCDs was above 90% at 100 mg/L	[29]

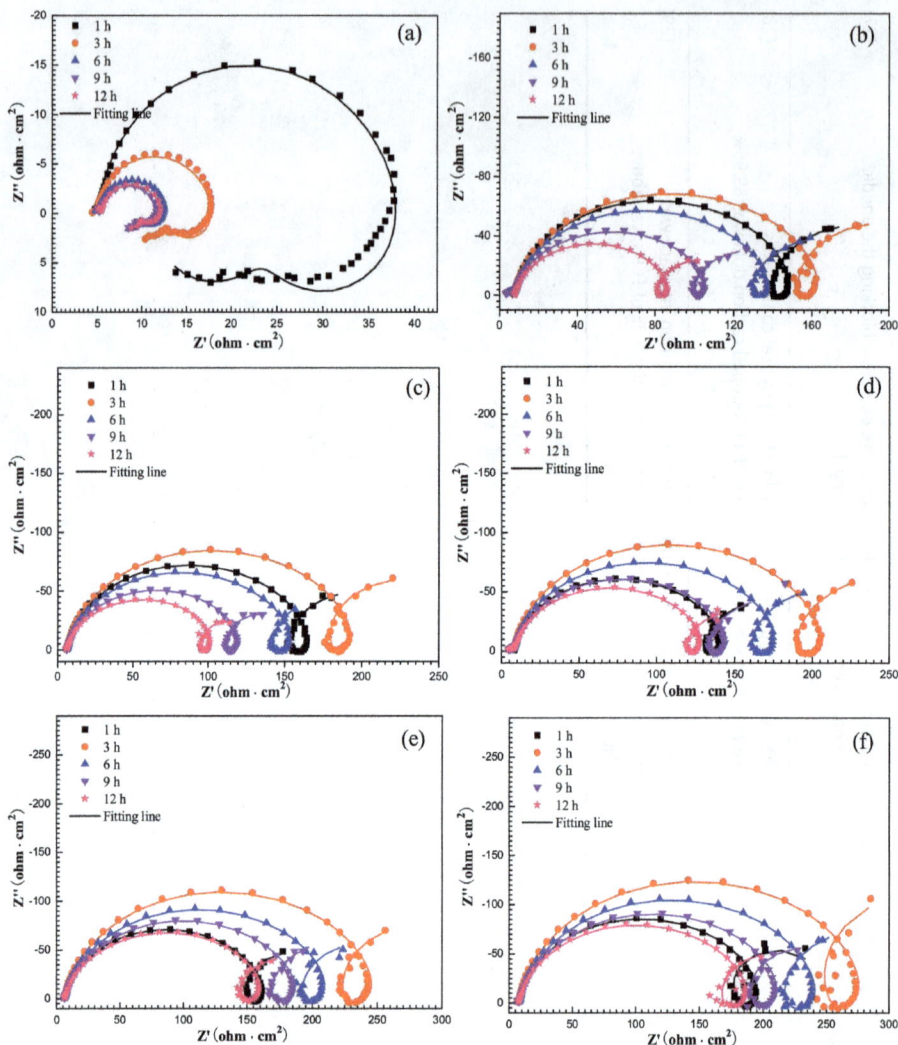

Figure 8.2: Nyquist plots for aluminum alloy measured in various immersion time with different concentrations of N, S-CDs for (a) blank, (b) 1 mg/L, (c) 2 mg/L, (d) 3 mg/L, (e) 4 mg/L and (f) 5 mg/L [26].

have a diameter and height of 3–5 nm. EIS is used to analyze the corrosion resistance performance of the coatings without and with N-CDs after 70 days of immersion in 3.5 wt% aqueous NaCl. The findings suggest that the bond interactions between N-CDs and polymer chains, the defect-repairing properties of N-CDs on iron surface coatings with 0.5 wt% N-CDs their better anticorrosive performance [26]. In addition, Figure 8.4 exhibits the SEM microphotographs of inhibitor at various concentrations.

An easy-to-use tool AFM is preferred to characterize the adsorbate with 3D distribution at metal/solution interfaces in corrosion processes. It is utilized to observe the

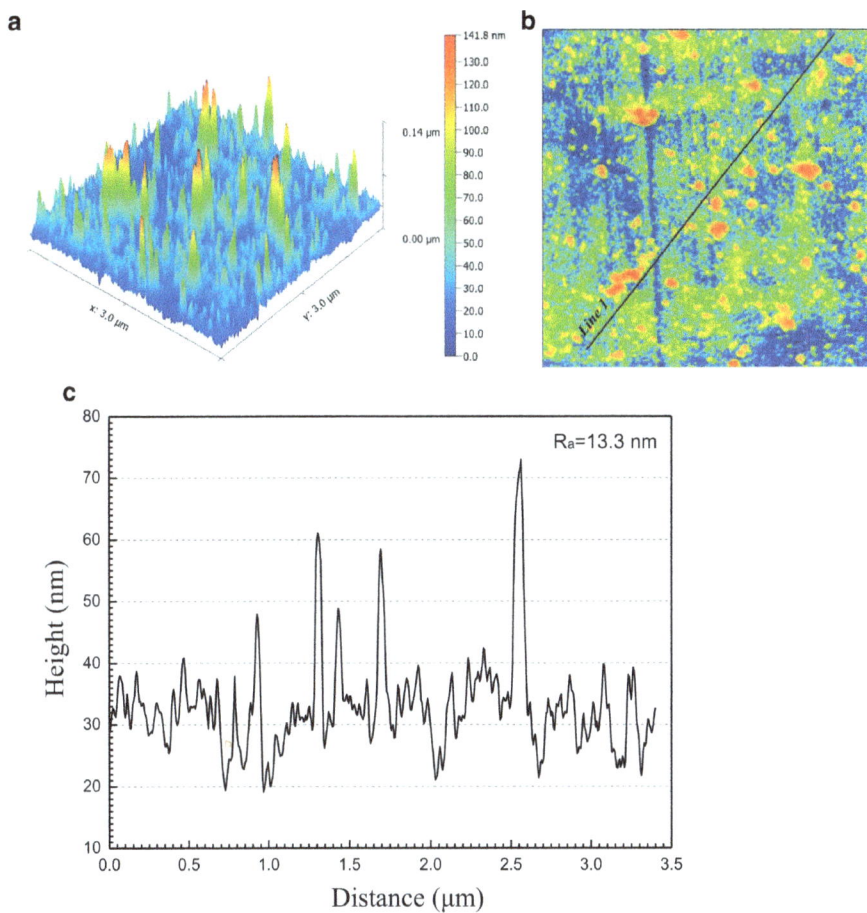

Figure 8.3: The surface morphology of carbon steel immersed with 50 mg/L N, S-CDs for (a) 3D AFM image, (b) 2D AFM image and (c) corresponding height profile graphs of Line 1 indicated in 2D image for 5 h at 50 °C [27].

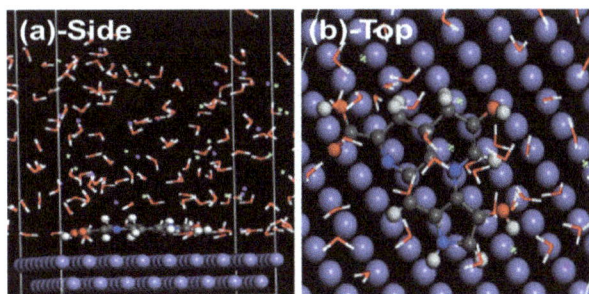

Figure 8.4: Equilibrium adsorption configurations of as-obtained N-CDs molecule on Fe (110) surface: (a) side and (b) top [27].

surface morphology under a microscope. The height profile graphs of the designated region in the 2D picture for carbon steel that was submerged in solution for 5 h with 50 mg/L of N, S-CDs are shown in Figure 8.5, along with the 2D AFM image that corresponds to the 3D-distributed image. The 3D AFM pattern (Figure 8.3) of the coupon displays the distribution and micromorphology of N, S-CDs, which formed a dense film with particles as tall as 141.8 nm in the viewing area. As a result, the inhibitory film made of N, S-CDs can efficiently and fully cover the surface of carbon steel [27]. Also Figure 8.4 exhibited the MD simulation images of N-CDs molecule on Fe (110) surface rom top and flat orientation.

8.4 Adsorption mechanism of CDs and heteroatom-doped CDs

The corrosion processes on the steel surface are primarily caused by hydrogen and anion. Numerous amino, hydroxyl and carboxyl functional groups found in the N-CDs made it simple for them to interact chemically with the metal surface and create coordinating bonds. Additionally, under the influence of hydrogen ions, some heteroatoms of

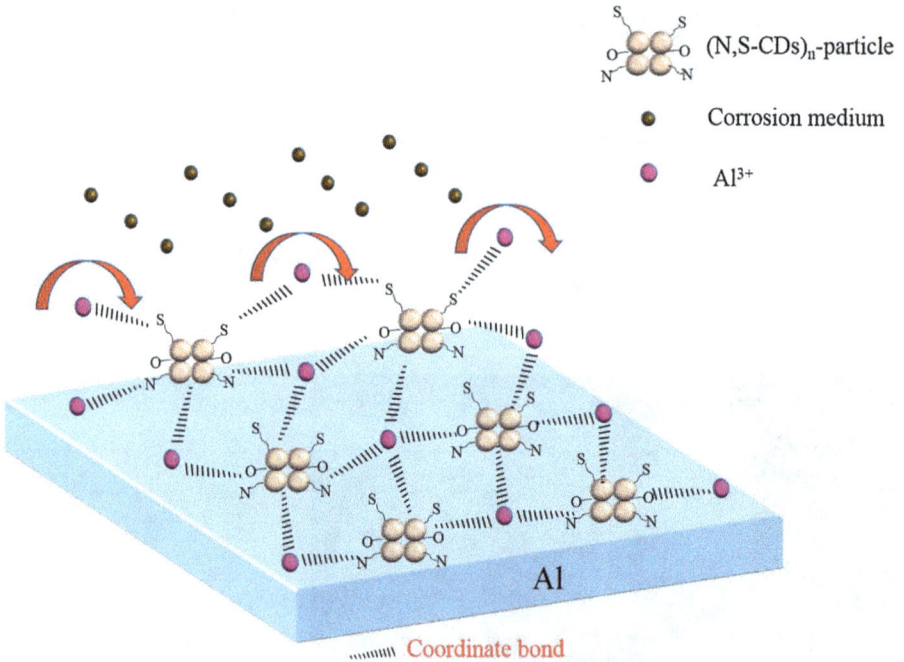

Figure 8.5: Schematic illustration of inhibition mechanism on aluminum surface for N, S-CDs in solution [26].

N-CDs were converted into the protonated form of N-CDs, which interacted electrostatically with the metal surface. Physical adsorption is a type of adsorption that can promote N-CDs' ability to inhibit. The inhibitor contains graphitic N, pyridine-N, pyrrole-like N and O atoms. The lone electron pairs of the pyrrole-like N and the pyridine-N help the corrosion inhibitor's chemical interactions. In order to create covalent connections and deposit inhibitor molecules as a protective layer on the metal surface, electron pairs are shared with iron's open 3d orbitals. This prevents additional aggressive anionic acid solution attack. For the majority of organic inhibitors, adsorption and coordination are thought to be the main inhibitory mechanisms. Only a few organic molecules containing O, N and S have been found to directly establish covalent bonds on the metal surface by chemisorption because the outer layer of aluminum lacks the d orbital [5]. The different chelating complexes of Al^{3+}, however, exhibit simple coordination with hydroxyl, carboxyl and heterocyclic. The addition of N, S-CDs and the catalyst inhibits both the hydrogen evolution processes and the dissolution of aluminum, as seen by the polarization curves in Figure 8.2. In addition, schematic illustration of inhibition mechanism on aluminum surface for N, S-CDs is shown in Figure 8.5.

8.5 Conclusion

In conclusion, CDs and heteroatom-doped CDs work wonders at preventing corrosion in corrosive solutions on metals. The electrochemical results demonstrated that the active sites at iron reduction were significantly hindered in the presence of inhibitors in acidic and saline media, demonstrating that the advanced materials effectively controlled both cathodic and anodic electrochemical reactions. Additionally, CDs offer great antibacterial capabilities, strong biocompatibility, low toxicity, chemical stability, high thermal activity and nonflammability. With these characteristics, CD-mediated inhibitors may soon dominate the corrosion inhibitor market. The endothermic nature of adsorption and inhibition was demonstrated by the standard enthalpy, whereas CD adsorption on the metal surface was spontaneous. According to Langmuir isotherms, the CDs were primarily mixed-type inhibitors.

Conflicts of interest: Authors declare no conflict of interests.

References

[1]   Laadam, G., El Faydy, M., Benhiba, F., Titi, A., Amegroud, H., Al-Gorair, A.S., Hawsawi, H., Touzani, R., Warad, I., Bellaouchou, A., Outstanding anti-corrosion performance of two pyrazole derivatives on carbon steel in acidic medium: Experimental and quantum-chemical examinations, Journal of Molecular Liquids, 2023, 121268.

[2]   Verma, D.K., Kazi, M., Alqahtani, M.S., Syed, R., Berdimurodov, E., Kaya, S., Salim, R., Asatkar, A., Haldhar, R., N–hydroxybenzothioamide derivatives as green and efficient corrosion inhibitors for mild steel: Experimental, DFT and MC simulation approach, Journal of Molecular Structure, 2021, 1241, 130648.

[3]   Qiang, Y., Guo, L., Li, H., Lan, X., Fabrication of environmentally friendly Losartan potassium film for corrosion inhibition of mild steel in HCl medium, Chemical Engineering Journal, 2021, 406, 126863.

[4]   Verma, D.K., Ebenso, E.E., Quraishi, M., Verma, C., Gravimetric, electrochemical surface and density functional theory study of acetohydroxamic and benzohydroxamic acids as corrosion inhibitors for copper in 1 M HCl, Results in Physics, 2019, 13, 102194.

[5]   Nofrizal, S., Rahim, A.A., Saad, B., Bothi Raja, P., Shah, A.M., Yahya, S., Elucidation of the corrosion inhibition of mild steel in 1.0 M HCl by catechin monomers from commercial green tea extracts, Metallurgical and Materials Transactions A, 2012, 43, 1382–1393.

[6]   Verma, D.K., Aslam, R., Aslam, J., Quraishi, M., Ebenso, E.E., Verma, C., Computational modeling: Theoretical predictive tools for designing of potential organic corrosion inhibitors, Journal of Molecular Structure, 2021, 1236, 130294.

[7]   Cao, S., Liu, D., Ding, H., Wang, J., Lu, H., Gui, J., Task-specific ionic liquids as corrosion inhibitors on carbon steel in 0.5 M HCl solution: An experimental and theoretical study, Corrosion Science, 2019, 153, 301–313.

[8]   Chaouiki, A., Lgaz, H., Chung, I.-M., Ali, I., Gaonkar, S.L., Bhat, K., Salghi, R., Oudda, H., Khan, M., Understanding corrosion inhibition of mild steel in acid medium by new benzonitriles: Insights from experimental and computational studies, Journal of Molecular Liquids, 2018, 266, 603–616.

[9]   Chong, A.L., Mardel, J.I., MacFarlane, D.R., Forsyth, M., Somers, A.E., Synergistic corrosion inhibition of mild steel in aqueous chloride solutions by an imidazolinium carboxylate salt, ACS Sustainable Chemistry & Engineering, 2016, 4, 1746–1755.

[10] Verma, C., Verma, D.K., Ebenso, E.E., Quraishi, M.A., Sulfur and phosphorus heteroatom-containing compounds as corrosion inhibitors: An overview, Heteroatom Chemistry, 2018, 29, e21437.

[11] KaHlmaHn, E., Routes to the development of low toxicity corrosion inhibitors for use in neutral solutions, A= Orking Party Report on Corrosion Inhibitors, 1994, 12.

[12] Sahu, S., Behera, B., Maiti, T.K., Mohapatra, S., Simple one-step synthesis of highly luminescent carbon dots from orange juice: Application as excellent bio-imaging agents, Chemical Communications, 2012, 48, 8835–8837.

[13] Dhenadhayalan, N., Lin, K.-C., Suresh, R., Ramamurthy, P., Unravelling the multiple emissive states in citric-acid-derived carbon dots, The Journal of Physical Chemistry C, 2016, 120, 1252–1261.

[14] De, B., Karak, N., A green and facile approach for the synthesis of water soluble fluorescent carbon dots from banana juice, Rsc Advances, 2013, 3, 8286–8290.

[15] Vasimalai, N., Vilas-Boas, V., Gallo, J., De Fátima Cerqueira, M., Menéndez-Miranda, M., Costa-Fernández, J.M., Diéguez, L., Espiña, B., Fernández-Argüelles, M.T., Green synthesis of fluorescent carbon dots from spices for in vitro imaging and tumour cell growth inhibition, Beilstein Journal of Nanotechnology, 2018, 9, 530–544.

[16] Zhang, J., Yu, S.-H., Carbon dots: Large-scale synthesis, sensing and bioimaging, Materials Today, 2016, 19, 382–393.

[17] Wang, F., Pang, S., Wang, L., Li, Q., Kreiter, M., Liu, C.-Y., One-step synthesis of highly luminescent carbon dots in noncoordinating solvents, Chemistry of Materials, 2010, 22, 4528–4530.

[18] Zheng, X.T., Ananthanarayanan, A., Luo, K.Q., Chen, P., Glowing graphene quantum dots and carbon dots: Properties, syntheses, and biological applications, Small, 2015, 11, 1620–1636.

[19] Liu, Z., Ye, Y., Chen, H., Corrosion inhibition behavior and mechanism of N-doped carbon dots for metal in acid environment, Journal of Cleaner Production, 2020, 270, 122458.

[20] Wang, J., Du, P., Zhao, H., Pu, J., Yu, C., Novel nitrogen doped carbon dots enhancing the anticorrosive performance of waterborne epoxy coatings, Nanoscale Advances, 2019, 1, 3443–3451.

[21] Luo, J., Cheng, X., Zhong, C., Chen, X., Ye, Y., Zhao, H., Chen, H., Effect of reaction parameters on the corrosion inhibition behavior of N-doped carbon dots for metal in 1 M HCl solution, Journal of Molecular Liquids, 2021, 338, 116783.
[22] Ye, Y., Zhang, D., Zou, Y., Zhao, H., Chen, H., A feasible method to improve the protection ability of metal by functionalized carbon dots as environment-friendly corrosion inhibitor, Journal of Cleaner Production, 2020, 264, 121682.
[23] Ye, Y., Yang, D., Chen, H., Guo, S., Yang, Q., Chen, L., Zhao, H., Wang, L., A high-efficiency corrosion inhibitor of N-doped citric acid-based carbon dots for mild steel in hydrochloric acid environment, Journal of Hazardous Materials, 2020, 381, 121019.
[24] Cui, M., Ren, S., Zhao, H., Wang, L., Xue, Q., Novel nitrogen doped carbon dots for corrosion inhibition of carbon steel in 1 M HCl solution, Applied Surface Science, 2018, 443, 145–156.
[25] Cao, S., Liu, D., Wang, T., Ma, A., Liu, C., Zhuang, X., Ding, H., Mamba, B.B., Gui, J., Nitrogen-doped carbon dots as high-effective inhibitors for carbon steel in acidic medium, Colloids and Surfaces A, Physicochemical and Engineering Aspects, 2021, 616, 126280.
[26] Cen, H., Zhang, X., Zhao, L., Chen, Z., Guo, X., Carbon dots as effective corrosion inhibitor for 5052 aluminium alloy in 0.1 M HCl solution, Corrosion Science, 2019, 161, 108197.
[27] Cen, H., Chen, Z., Guo, X., N, S co-doped carbon dots as effective corrosion inhibitor for carbon steel in CO2-saturated 3.5% NaCl solution, Journal of the Taiwan Institute of Chemical Engineers, 2019, 99, 224–238.
[28] Ye, Y., Jiang, Z., Zou, Y., Chen, H., Guo, S., Yang, Q., Chen, L., Evaluation of the inhibition behavior of carbon dots on carbon steel in HCl and NaCl solutions, Journal of Materials Science & Technology, 2020, 43, 144–153.
[29] Ye, Y., Yang, D., Chen, H., A green and effective corrosion inhibitor of functionalized carbon dots, Journal of Materials Science & Technology, 2019, 35, 2243–2253.

Muhammed Safa Çelik, Hüseyin Fatih Çetinkaya, Serap Çetinkaya,
Gamze Tüzün, Burak Tüzün*

Chapter 9
Polymeric nanoparticles and their composites in corrosion inhibition

Abstract: Metal hardware is inherently prone to corrosion. Substantial research has been undertaken to explore to supersede this challenge. Initially, paints, pigments and organic coatings were utilized to circumvent metal corrosion. Lately, polymer composites and nanocomposites have become favored anticorrosion agents, especially epoxy, polyethylene glycol, polyaniline and polystyrene. The future research on polymer nanocomposites has the potential to address the current challenges of metal corrosion.

Keywords: Corrosion, corrosion inhibitors, nanoparticles, nanopolymers

9.1 Introduction

Today, nanotechnology is a swift progress in Garea. Nanoparticles are between 1 and 100 nm in size due to their exclusive physical and chemical properties. The most important field of application of nanoparticles in industry is that they have a high specificity to protect metals against corrosion in adverse conditions [1, 2]. Metals are indispensable for the industrial life because they are electrical and thermal conductors and have desirable freezing and boiling points, tensile strengths, high mass/volume ratios and plastic behavior. The omnipresence of free oxygen is the noxious enemy for these precious elements, giving rise to corrosion [3, 4]. Various techniques are used to reduce the corrosion rate such as the coating of metal surfaces and the use of corrosion inhibitors [5]. The inhibitors are applied on the metal surface as thin films to insulate it from its environment. They can be organic and inorganic in chemistry. A common drawback in the use of these expensive substances is that they are

---

*Corresponding author: Burak Tüzün,** Plant and Animal Production Department, Technical Sciences Vocational School of Sivas, Sivas Cumhuriyet University, Sivas, Turkey,
e-mail: theburaktuzun@yahoo.com, http://orcid.org/0000-0002-0420-2043
Muhammed Safa Çelik, Serap Çetinkaya, Department of Molecular Biology and Genetics, Science Faculty, Sivas Cumhuriyet University, Sivas 58140, Turkey
Hüseyin Fatih Çetinkaya, Department of Environmental Engineering, Faculty of Engineering, Sivas Cumhuriyet University, Sivas, Turkey
Gamze Tüzün, Department of Chemistry, Faculty of Science, Sivas Cumhuriyet University, Sivas, Turkey

https://doi.org/10.1515/9783111071756-009

detrimental to human health. In this regard, nanoparticles and nanocomposites stand as better corrosion inhibitors as they are relatively more harmless, biodegradable and less expensive [6–8].

9.1.1 Properties of corrosion

9.1.1.1 Corrosion definition

All materials, or more generally all kinds of products and structures made using materials, are generally exposed to physical wear during their use. The underlying cause of this wear can be mechanical, thermal, chemical, electrochemical or microbiological. In terms of metals, the main reason for this wear problem is corrosion. According to DIN EN ISO 8044, corrosion is defined as the significant loss of function of metals, the environment or the technical systems of which they are a part, as a result of the physicochemical interaction of metals with their environment [9]. This interaction is usually electrochemical [10].

The main reason why metals can easily corrode even in normal operating environments is that metals irresistibly tend to return to their stable state. Almost all of the metals exist in nature as low-energy oxide compounds, that is, in a stable state. The foundation of the corrosion problem was laid with the first examples of human beings being able to obtain them by processing metal mines in nature, thanks to metal mining, which dates back to prehistoric times. Large amounts of heat energy are transferred to the metal oxides extracted from the mines to separate them from the oxygen in the blast furnaces, forcing the metals to be in a thermodynamically unstable state. This transferred excess energy is the driving force in the initiation of various corrosion reactions later on. When the corrosion reaction is complete, excess energy is released and the metal (for example, iron and steel) returns to its stable, i.e. oxidized state, completing the cycle shown in Figure 9.1 [10, 11].

It is thought that the economic loss caused by the corrosion of metals is at the level of trillions of dollars worldwide and it is more than 3% of the world's gross national product, namely GDP [12]. In addition to this huge economic cost, corrosion poses a serious risk for both humans and the environment. The destruction of thousands of houses and the death of hundreds of people in the explosion that took place due to direct corrosion in Guadalajara in 1992 is one of the biggest examples that reveal the magnitude of this risk. As a result of the leakage that occurred in the transit pipeline belonging to BP in Prudhoe Bay in 2006, the spillage of approximately 267,000 gallons of crude oil into the environment for five days also revealed the magnitude of the environmental risk that corrosion may cause. As can be understood from these two examples, it is of great importance to develop new products and techniques that can prevent or slow down corrosion, considering the accidents that corrosion may cause and the huge economic loss [10].

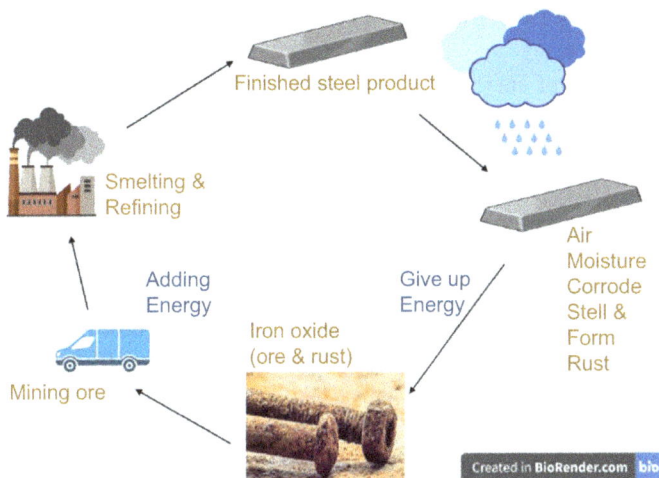

Figure 9.1: Corrosion cycle in metals.

9.1.1.2 Types of corrosion

In general, corrosion can be examined in three main groups according to the detection methods [10]:

Corrosion types detectable by visual inspection
- Uniform corrosion: A type of corrosion in which metal loss occurs at almost the same rate across the entire surface.
- Local corrosion (pitting corrosion, crevice corrosion): A type of corrosion in which metal loss occurs in certain areas.
- Galvanic corrosion: A type of corrosion caused by electrical contact between dissimilar conductors in an electrolyte.

Types of corrosion that can be detected using special inspection instruments
- Intergranular corrosion (stripping corrosion): The type of corrosion that takes place at the grain boundaries in the metal structure.
- Selective corrosion: A type of corrosion that results from the selective dissolution of one or more of the components in an alloy.
- Velocity-acting corrosion (erosion–corrosion, cavitation and erosion): Erosion–corrosion caused by high velocity flow, cavitation occurring at higher velocity flow, injury caused by vibrational movement of the surface with the surface in close contact under load.

Corrosion types that can be detected with the help of a microscope
- Cracking event (stress corrosion, cracking, fatigue): Corrosion types that cause mechanical events.

– High temperature corrosion (internal attacks, calcification): Types of corrosion caused by high temperature.
– Microbial corrosion: Types of corrosion caused by certain types of bacteria or microbes.

9.1.1.3 Corrosion measurement techniques

The methods used in the measurement of corrosion behavior can be broadly divided into two main groups: electrochemical and nonelectrochemical techniques. Nonelectrochemical techniques are mass loss, pitting and crack formation rate, surface measurements, analytical methods and mechanical testing. Mass loss tests are the simplest and most widely used corrosion and inhibitor test method and are used to monitor total metal thinning and local corrosion types as a function of inhibitor concentration. The corrosion test-strip method follows the principle of monitoring the corrosion development by visual, microscopic or mass loss methods by exposing test strips of a specified standard weight, size and shape to a corrosive environment with and without inhibitor for 14-day period. With this method, although the safe lifetime can be predicted depending on the rate of corrosion, it can provide little information about the specific events that cause the total corrosion damage in the specified period [10].

Since corrosion is basically a process involving electrochemical oxidation and reduction reactions, it is much more practical to use electrochemical measurement methods for corrosion tests, but they can also provide more detailed information [13]. Electrochemical measurement techniques used in corrosion tests are potentiodynamic polarization measurements, electrochemical impedance measurements, electrochemical noise measurement, scanning kelvin probe and scanning-vibrating electrode technique [10].

9.1.1.4 Corrosion cost and outcome

Corrosion is a worldwide phenomenon and leads to the 3.4% loss in world's income. Thus, direct and indirect loss amounts to US $2.5 trillion [14]. Direct losses arise from the damaged or faulty machinery infrastructure and the cost of maintenance [15, 16].

One detrimental outcome of corrosion is the inadvertent leakage of contaminated liquids from transporting vessels or storage tanks. In 2011, earthquakes and tsunami caused the reactors to melt down [17], and in 2013, hundreds of tons of wastewater leaked from a storage tank because of the corrosion around the faulty seals in Japan [4]. On November 22, 2013, the Donghuang II oil pipeline blew up by the ignition of vaporous oil in eastern China [18–20].

9.1.1.5 Corrosion protection methods

It has been estimated that application of the best metal corrosion prevention measures may reduce annual corrosion costs up to 35% [21]. Various strategies have been adopted to combat the menace of corrosion: (i) designing a material or system to avoid cracks, excessive speeds or local turbulence [22]; (ii) selection of corrosion resistant materials for specific applications; (iii) electrochemical (anodic and cathodic protection); (iv) coatings; and (v) use of corrosion inhibitors. Although no metal is completely immune to corrosion, metals react differently to corrosion in different environments, so by controlling and understanding the application environment, metals with good corrosion resistance can be selected [23]. Anodic protection is employed to shield protect carbon steels in extreme pH environments [24]. Coatings (painting, coating and lubrication) involve covering a metal surface with a protective film that forms a physical barrier to corrosive surroundings [25]. Corrosion inhibitors are compounds that are added in small quantities to destructive environs to lessen the level of corrosion [22]. Corrosion inhibitors are classified by their inhibition mechanism (Figure 9.2). They are liquid or vapor films at the metal/environment interface [20, 22].

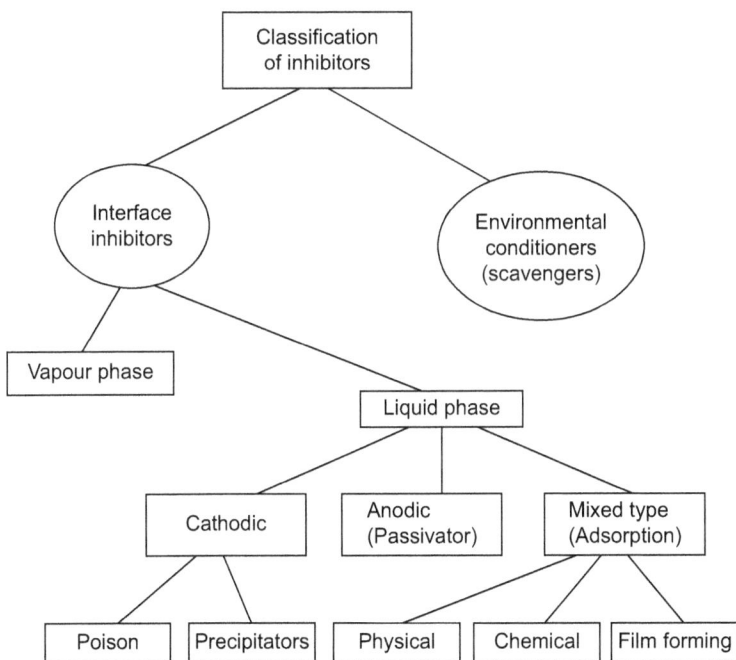

Figure 9.2: Classification of inhibitors [20, 22].

9.1.2 Nanomaterials as corrosion inhibitors

9.1.2.1 Nanoparticles

Shifting the focus of the search from organic coatings to the use of nanomaterials has been prompted by the proposition that the smaller size of the components of nanomaterials will increase adhesion to metal surfaces. Research in this field has mainly driven by factors as follows:
- Negative opinions on the link between hexavalent chromium and carcinogenicity (Figure 9.3) [26]
- Hazardous volatile organic compounds
- The sufficiency of very small amounts of nanomaterials to obtain a thin, sticky film

The issue of the link between hexavalent chromium and carcinogen etic	Due to hydrolysis Cr^{6+} exists in moderate acidic conditions as an oxo ion
	Hydrolysis of Cr^{6+} has an impact on corrosion protection/inhibition
	Damage to DNA is not due to Cr^{6+} or Cr^{3+} but by the "debris" resulting from the conversion of Cr^{6+} to Cr^{3+}
	CrO_4^{2-} "aşone" is not responsible for the damage of DNA
	Inhibition of corrosion by chromates could be attributed to irreversible adsorption of chromates on metals and metal oxide surfaces

Figure 9.3: Alleged link between hexavalent chromium and cancer [26].

Ultrathin silane films including small amounts of nanoparticles have been shown to provide exceptional protection to metals [27]. The procedure involved the following steps: colloidal solutions were obtained by loading silica nanoparticles into bisulfur silane. After the metal was then immersed in this solution and dried, a very thin film formed on the metal surface. In another instance, CeO nanoparticles loaded in silica–alumina hybrid coating have been successful in the inhibition of Cu corrosion [28]. In another application, electrochemical performance of the cerium–silane hybrid coating has been shown to rely on the concentration of ceria nanoparticles. This information illuminated the significant effect of nanoparticle concentration on the barrier properties of silane films [29].

An electroless technique has been chosen to deposit Ni–phosphorus (P) and Ni–P–Re on copper disks on metallic surfaces [30]. This Ni-containing film has been found to be helpful in automobile and electronics industries. Corrosion resistance has been further enhanced by adding nano-S_iO_2 particles into the Ni–phosphorus coating [31]. In another application, brass surface has been coated with monolayers of PropS-

SH (3-mercapto-propyl trimethyloxy) silane tempered with La_2O_3 nanoparticles [32]. The latter filled the holes in the silane film and increased the density of the film. Conversion of some oxide nanoparticles to hydroxide caused blocking of the cathode sites by the resulting hydroxides [33]. In this way, it was possible to effectively prevent the access of corrosive substances to the metal surface. Clay nanoparticles have been combined with cerium salt to increase the corrosion resistance of the mild steel coating due to the synergistic effect between the nanoparticles and the clay [33]. In addition, both corrosion inhibitions and antibacterial properties of Ag nanoparticles have been explored for aluminum in HCl solution [34]. Ag nanoparticles have effectively inhibited corrosion in both cases [26, 34].

9.1.2.2 Nanopolymers

Corrosion of metals is an electrochemical reaction with the surrounding milieu, where the metal behaves as the anode and oxidized (Figure 9.1) and the oxygen existing along with water is the cathode, producing hydroxyl ions [1, 2]. Corrosion of metals and metal alloys is also encouraged thermodynamically. Nonconductive polymers like epoxies have superior compressive strengths, and polymers reinforced with nanoparticles have been used as the corrosion-protecting agents. Nanosilica-filled epoxy, for example, improves durability and toughness and yields high thermal stability [35]. Polymers such as polyaniline (PANI), polythiophene, polypyrrole and polyacetylene show conductive and anticorrosive properties, depending on their oxidation state [36]. Here, the self-healing ability of polymers appears to play an important role in protecting metals. Conductive polymers can also prevent oxide formation. PANI stabilizes the passive state of stainless steel against sulfuric acid [37]. PANI/partially phosphorylated poly(vinyl alcohol) nanoparticles in the epoxy matrix have displayed good corrosion resistances [38–40].

9.1.2.3 Nanocomposite as corrosion inhibitor

Nanocomposites hold another means to tackle corrosion. Polymers and nanomaterials are used to produce nanocomposites. In general, organic and inorganic components constitute a nanocomposite [34]. The inorganic components of the nanocomposite provide its tackiness, high ductility and good mechanical rigidity, while the organic components contribute to its flexibility and reduce structural defects. The commonest polymers in the preparation of nanocomposite coatings are epoxy [41, 42], polyurethane [43], polyethylene glycol (PEG) [44], PANI [45], polystyrene [46], polyacrylic [47], polyvinyl alcohol [48], polypyrrole [49] and polyamide [50]. Metal nanoparticles [45, 51] and their oxides [52], carbides [53] and phosphate [54] generally form the inorganic component [8].

Carbon nanoparticles (fullerene, nanodiamond, graphene, graphene oxide, carbon nanotube, carbon black, nanoclay, silica and titania nanoparticles) have been incorporated into matrices to increase the corrosion resistance of polymeric nanocomposites. To improve the performance of nanocomposites, functionalization is adopted to achieve better interfacial bonding and charge transfer properties of the matrix/nano filler [55]. The addition of functional nanoparticles, besides preventing the diffusion of penetrating species, strengthens the nanocomposite structure [56]. This process can also ease friction and increase wear resistance of the base material [40, 57, 58].

9.1.3 Conclusions

Metals have desired inherent characteristics like high electrical conductivity, strength, robustness and toughness. This is why metal is preferred in various structural, engineering, electrical and electronic uses. However, metals have a tendency to react with surrounding oxygen and rust. The decomposing environs usually determine the magnitude of corrosion and the underlying mechanisms. Corrosion affects mechanical strength and causes environmental pollution at the same time [47]. Polymers can be exploited as corrosion inhibitors, although their effectiveness is inadequate [48]. In order to produce new anticorrosion polymer nanocomposites in advanced engineering uses, various aspects of these materials must be investigated. The corrosion resistance of nanocomposite coatings essentially depends on the distribution of the nanofiller in the matrix, and a homogeneous dispersion is often a difficult task. Here, in situ polymerization and rapid sonication methods can be convenient.

Temperature and environmental factors directly affect the corrosion resistance of coatings. Nanocomposite coatings endure temperature vagaries and aging better than pure polymers. The incorporation of functionalized nanoparticles can enhance the interaction between matrix/nanophils [49]. It is also important to understand the corrosion inhibition mechanism and adhesion properties of a particular nanocomposite [40, 50, 51]. The adsorption energies of corrosion inhibitors need to be studied by means of thermodynamic and molecular dynamics simulations. Nanocomposites also need to be investigated for nonwetting, surface roughness and anticorrosion properties using contact angle, salt spray testing and scale prevention. Future aerospace industry entails new stratagems to gain insight into the corrosion mechanism.

For future fuel cells, graphene-based nanocomposite coating can increase bipolar plate hydrophobicity and replace chromium and other potentially toxic chemicals. In addition, polymer/CNT and polymer/graphene have excellent corrosion resistance and durability properties for use in next-generation fuel-cell bipolar plates. Advanced nanocomposite can serve as better alternatives to current artificial bones and tissues. Current consciousness has opened new horizons in corrosion research to adopt green nanomaterials.

References

[1] Hameed, R.S., et al., Nano-composite as corrosion inhibitors for steel alloys in different corrosive media, Advances in Applied Science Research, 2013, 4(3), 126–129.

[2] Khodair, Z.T., et al., Corrosion protection of mild steel in different aqueous media via epoxy/nanomaterial coating: Preparation, characterization and mathematical views, Journal of Materials Research and Technology, 2019, 8(1), 424–435.

[3] Umoren, S.A., Solomon, M.M., Recent developments on the use of polymers as corrosion inhibitors-a review, The Open Materials Science Journal, 2014, 8(1).

[4] Dwivedi, D., et al., Carbon steel corrosion: A review of key surface properties and characterization methods, RSC Advances, 2017, 7(8), 4580–4610.

[5] Fotovvati, B., et al., On coating techniques for surface protection: A review, Journal of Manufacturing and Materials Processing, 2019, 3(1), 28.

[6] Abdeen, D.H., et al., A review on the corrosion behaviour of nanocoatings on metallic substrates, Materials, 2019, 12(2), 210.

[7] Rathish, R.J., et al., Corrosion resistance of nanoparticle-incorporated nano coatings, European Chemical Bulletin, 2013, 2(12), 965–970.

[8] Jain, P., et al., Jeetendra. Potential of nanoparticles as a corrosion inhibitor: A review, Journal of Bio- and Tribo-Corrosion, 2020, 6(2), 1–12.

[9] Corrosion of metals and alloys – Basic terms and definitions (ISO 8044:2015); Trilingual version EN ISO 8044:2015. publication date: 2015-07-01; ics: 77.060, 01.040.77.

[10] Topal, E., Quantum chemical investigation of some proton pump inhibitors used as corrosion inhibitors for aluminum. 2016. Master's Thesis. Bursa Technical University.

[11] Sahan, Y., et al. Performance of the electrical generator cell by the ferrous alloys of printed circuit board scrap and Iron Metal 1020. In: IOP Conference Series: Materials Science and Engineering. IOP Publishing, 2018, 012038.

[12] Thompson, N.G., et al., Cost of corrosion and corrosion maintenance strategies, Corrosion Reviews, 2007, 25(3–4), 247–262.

[13] Princeton Applied Research, Electrochemistry and Corrosion: Overview and Techniques, Application Note, CORR-4, EG&G Princeton Applied Research Corp., USA, 1997.

[14] Hou, B., et al., The cost of corrosion in china., Npj Materials Degradation, 2017, 1(1), 1–10.

[15] Revie, R.W., Corrosion and Corrosion Control: An Introduction to Corrosion Science and Engineering, John Wiley & Sons, 2008.

[16] Elayaperumal, K., et al., Corrosion Failures: Theory, Case Studies, and Solutions, John Wiley & Sons, 2015.

[17] Seth, D., Fukushima Cleanup Could Drag on for Decades. CBSNews. CBS Interactive, 27 Jan. 2014.

[18] Agence France Presse. Sinopec to pay compensation over pipeline blast. Channel News Asia MediaCorp Pte Ltd, 13 Jan. 2014.

[19] Wayne, M., Tejada, C., China Cites Lapses in Sinopec Pipeline Blasts, Wall Street J Dow Jones & Company, 2014, 9.

[20] Umoren, S.A., Solomon, M.M., Protective polymeric films for industrial substrates: A critical review on past and recent applications with conducting polymers and polymer composites/nanocomposites, Progress in Materials Science, 2019, 104, 380–450.

[21] Koch, G.H., et al., Corrosion Cost and Preventive Strategies in the United States, Federal Highway Administration, United States, 2002.

[22] Umoren, S.A., Solomon, M.M., Recent developments on the use of polymers as corrosion inhibitors-a review, The Open Materials Science Journal, 2014, 8(1).

[23] Baekmann, W., Schwenk, W., Fundamentals and Practice of Electrical Measurements, in Handbook of Cathodic Corrosion Protection, Gulf Professional Publishing, 1997, 79–138.

[24]   Edeleanu, C., Corrosion control by anodic protection, Platinum Metals Review, 1960, 4(3), 86–91.

[25]   Smith, L., Control of corrosion in oil and gas production tubing, British Corrosion Journal, 1999, 34 (4), 247–253.

[26]   Moodley, G.K., Advances in corrosion inhibition materials and technologies: A review, Advanced Materials Letters, 2019, 10(4), 231–247.

[27]   Palanıvel, V., et al., Nanoparticle-filled silane films as chromate replacements for aluminum alloys, Progress in Organic Coatings, 2003, 47(3–4), 384–392.

[28]   Lakshmı, R.V., et al., Ceria nanoparticles vis-à-vis cerium nitrate as corrosion inhibitors for silica-alumina hybrid sol-gel coating, Applied Surface Science, 2017, 393, 397–404.

[29]   Zand, R.Z., et al., Effects of ceria nanoparticle concentrations on the morphology and corrosion resistance of cerium–silane hybrid coatings on electro-galvanized steel substrates, Materials Chemistry and Physics, 2014, 145(3), 450–460.

[30]   Wojevoda-Budka, J., et al., Microstructure characteristics and phase transformations of the Ni-P and Ni-P-Re electroless deposited coatings after heat treatment, Electrochimica Acta, 2016, 209, 183–191.

[31]   Gao, Z., et al., Corrosion behavior and wear resistance characteristics of electroless Ni–P–CNTs plating on carbon steel, International Journal of Electrochemical Science, 2015, 10(15), 637–648.

[32]   Fan, H.O., et al., Self-assembled (3-mercaptopropyl) trimethoxylsilane film modified with La2O3 nanoparticles for brass corrosion protection in NaCl solution, Journal of Alloys and Compounds, 2017, 702, 60–67.

[33]   Santana, I., et al., Hybrid sol-gel coatings containing clay nanoparticles for corrosion protection of mild steel, Electrochimica Acta, 2016, 203, 396–403.

[34]   Fetouh, H.A., et al., An electrochemical investigation in the anticorrosive properties of silver nanoparticles for the acidic corrosion of aluminium, Journal of Electrochemistry, 2018, 24(1), 89.

[35]   Conradi, M., et al., Mechanical and anticorrosion properties of nanosilica-filled epoxy-resin composite coatings, Applied Surface Science, 2014, 292, 432–437.

[36]   Ahmad, N., et al., Inhibition of corrosion of steels with the exploitation of conducting polymers, Synthetic Metals, 1996, 78(2), 103–110.

[37]   Sazou, D., Georgolıous, C., Formation of conducting polyaniline coatings on iron surfaces by electropolymerization of aniline in aqueous solutions, Journal of Electroanalytical Chemistry, 1997, 429(1–2), 81–93.

[38]   Chen, F., Liu, P., Conducting polyaniline nanoparticles and their dispersion for waterborne corrosion protection coatings, ACS Applied Materials and Interfaces, 2011, 3(7), 2694–2702.

[39]   Sababı, M., et al., Influence of polyaniline and ceria nanoparticle additives on corrosion protection of a UV-cure coating on carbon steel, Corrosion Science, 2014, 84, 189–197.

[40]   Kausar, A., Corrosion prevention prospects of polymeric nanocomposites: A review, Journal of Plastic Film & Sheeting, 2019, 35(2), 181–202.

[41]   Kım, H., et al., Enhancement of barrier properties by wet coating of epoxy-ZrP nanocomposites on various inorganic layers, Progress in Organic Coatings, 2017, 108, 25–29.

[42]   Pourhashem, S., et al., Exploring corrosion protection properties of solvent based epoxy-graphene oxide nanocomposite coatings on mild steel, Corrosion Science, 2017, 115, 78–92.

[43]   Huang, T.C., et al., Advanced anti-corrosion coatings prepared from α-zirconium phosphate/ polyurethane nanocomposites, RSC Advances, 2017, 7(16), 9908–9913.

[44]   Gu, M., et al., Cellulose nanocrystal/poly (ethylene glycol) composite as an iridescent coating on polymer substrates: Structure-color and interface adhesion, ACS Applied Materials and Interfaces, 2016, 8(47), 32565–32573.

[45]   Essıen, E.A., et al., Synthesis, characterization and anticorrosion property of olive leaves extract-titanium nanoparticles composite, Journal of Adhesion Science & Technology, 2018, 32(16), 1773–1794.

[46] Al Juhaıman, L.A., et al., Polystyrene/organoclay nanocomposites as anticorrosive coatings of C-steel, International Journal of Electrochemical Science, 2016, 11, 5618–5630.
[47] Sajjadı, S.A., et al., A comparative study on the effect of type of reinforcement on the scratch behavior of a polyacrylic-based nanocomposite coating, Journal of Coatings Technology and Research, 2013, 10(2), 255–261.
[48] Srımathı, M., et al., Polyvinyl alcohol–sulphanilic acid water soluble composite as corrosion inhibitor for mild steel in hydrochloric acid medium, Arabian Journal of Chemistry, 2014, 7(5), 647–656.
[49] Mahmoud, A.H., et al., Electrodeposition and corrosion protection performance of polypyrrole composites on aluminum, International Journal of Electrochemical Science, 2016, 11, 3938–3951.
[50] Ramezanzadeh, B., et al., A study on the anticorrosion performance of the epoxy–polyamide nanocomposites containing ZnO nanoparticles, Progress in Organic Coatings, 2011, 72(3), 410–422.
[51] Solomon, M.M., et al., Gum arabicsilver nanoparticles composite as a green anticorrosive formulation for steel corrosion in strong acid media, Carbohydrate Polymers, 2018, 181, 43–55.
[52] Quadri, T.W., et al., Zinc oxide nanocomposites of selected polymers: Synthesis, characterization, and corrosion inhibition studies on mild steel in HCl solution, ACS Omega, 2017, 2(11), 8421–8437.
[53] Racz, A.S., et al., Corrosion resistance of nanosized silicon carbide-rich composite coatings produced by noble gas ion mixing, ACS Applied Materials & Interfaces, 2017, 9(51), 44892–44899.
[54] Deshpande, P.P., et al., Conducting polyaniline/nano-zinc phosphate composite as a pigment for corrosion protection of low-carbon steel, Chemical Papers, 2016, 71(2), 189–197.
[55] Quaresimin, M., Salviato, M., Zappalorto, M., Strategies for the assessment of nanocomposite mechanical properties, Composites Part B: Engineering, 2012, 43, 2290–2297.
[56] Awaja, F., et al., Cracks, microcracks and fracture in polymer structures: Formation, detection, autonomic repair, Progress in Materials Science, 2016, 83, 536–573.
[57] Xu, B.S., et al., Progress of nano-surface engineering, International Journal Materials and Product Technology, 2003, 18, 338–346.
[58] Min, C., et al., Preparation and tribological properties of polyimide/carbon sphere microcomposite films under seawater condition, Tribilogy International, 2015, 90, 175–184.

Amir Hossein Jafari Mofidabadi, Nariman Alipanah, Ali Dehghani*

Chapter 10
Organic–inorganic hybrid nanostructured materials in corrosion prevention

Abstract: Regarding the contemporary industrial obstacles, the demand of new class of anticorrosion inhibitors emergence becomes more sensitive. Lack of substantial protection of the newly applied inhibitors and environmental hazard awareness of traditional inhibitors are good reasons for researchers to go through hybrid inhibitors. With the straightforward strategy, the first step was the combination of organic materials with other inhibitive compounds (e.g., other organics, metals or halides), which is a successful concept toward achieving hybrid self-assemble inhibitors. However, by entering in nano era, the expectation of control release capability alongside of sufficient anticorrosion impact results in speaking of MOFs application as smart coordinated hybrid inhibitors for corrosion prevention.

Herein, as a brief overview, the steps and progress made from first class of hybrid inhibitors toward generation of smart hybrid inhibitors is provided and discussed.

Keywords: Hybrid inhibitor, MOF, organic inhibitor, control-release, smart inhibitor

10.1 Introduction

Corrosion prevention now has been one of the important research fields where researchers focused on recently. To reach this goal, different methods are proposed. But among all, corrosion inhibitors application is the most recommended technique. Contemporary researches classified the protective agents in three main subgroups: organic, inorganic and mixed.

---

*Corresponding author: Ali Dehghani, Department of Chemical Engineering, Faculty of Engineering, Golestan University, Aliabad Katoul, Iran, e-mail: dehghaniali1996@gmail.com
Amir Hossein Jafari Mofidabadi, Department of Chemical Engineering, Faculty of Engineering, Golestan University, Aliabad Katoul, Iran
Nariman Alipanah, Department of Surface Coatings and Corrosion, Institute for Color Science and Technology (ICST), P.O. 16765654, Tehran, Iran

https://doi.org/10.1515/9783111071756-010

10.1.1 Application of inorganic materials and their limits

Inorganic inhibitors are the most traditional anticorrosion structures which have utilized for many years. They are slightly soluble in water. So they are partially dissolved in the solutions and release a certain amount of inhibitive agent that will be absorbed on the active areas of the metal substrate (anodic or cathodic). Inorganic inhibitors often prevent corrosion reactions by oxidizing the metal surface and forming a film on the active parts. A film, which is actually a passive layer, is constructed on these active areas and the metal is protected against corrosive agents [1–4]. For instance, zinc chromate is one of the most common anticorrosion inorganic pigments, which shows excellent performance in corrosion protection and emergence of self-healing ability in coatings. Chromate ion oxidizes other molecules as an oxidizing agent to create a protective layer on the surface of the film and is reduced to Cr_2O_3 or $Cr(OH)_3$. However, in recent years, the application of this material has been prevented due to its toxicity and serious environmental problems [2, 5–7]. This pigment was replaced by zinc phosphate as a well-performance inhibitive inorganic compound that is cheaper and more environmentally friendly than zinc chromate, although the use of zinc phosphate was also challenging due to its low solubility compared to zinc chromate in corrosive solution environments. Low solubility in saline solution reduces the ability to release active inhibitive agents in the electrolyte, resulting in weak active-inhibition appearance. It almost only had the role of a barrier property rather than active protection. In order to improve the inhibition properties of phosphate-based materials, second and third generation of inhibitive compounds were produced [2, 5–8].

As a solution for the solubility issue, researchers proposed some interesting strategies. Surely, physical and chemical modifications in the cationic and anionic parts of the low soluble materials can improve their solubility in corrosive solutions. Molybdenum phosphate and sodium zinc phosphate are examples of second-generation inhibitors, and zinc aluminum polyphosphate and zinc tripolyphosphate are also examples of third generation. Zinc aluminum phosphate, zinc potassium phosphate, zinc molybdenum phosphate and zinc polyphosphate are other popular pigments in these two generations. Although the increase in solubility led to the improvement of the protective properties, the systems containing these pigments still reflected a good protective behavior in a short time, corresponding to limited release of inhibitive substances in the electrolyte [2, 6–8].

10.1.2 Time to apply organic compounds as replacement

On the other hand, the organic inhibitive compounds were proposed following the research of researchers to find out a suitable alternative for inorganic compounds. Many organic inhibitors, such as benzimidazole and benzotriazole, have oxygen or sulfur or nitrogen heteroatoms which have lone pair electrons and can be adsorbed to the cations on the metal surface. As a result of the chemical absorption of organic compounds on the metal surface and the chelating of metal-organic connection, the metal substrate is protected against corrosive agents such as chloride ions. Organic inhibitors have high inhibitory power especially in acidic environment. There were two drawbacks in using organic inhibitors. First, this type of inhibitors shows weak inhibition effect at about neutral pH. Although remarkable inclusion of organic inhibitors seems a straightforward approach, achievements from literature provide some interesting information. Actually, at higher concentrations than optimum, no remarkable change would be observed. In other words, incorporation of high concentration of inhibitor is not a constructive way and researchers must follow other techniques for further corrosion prevention.

The second problem is related to the direct addition of organic inhibitors to the coatings. In coating phase, the interactions between inhibitor heteroatoms and functional groups of the polymer during the curing reaction reduce the cross-linking density and affect the barrier properties of the coating [1, 2, 6, 7, 9].

10.1.3 Emergence of hybrid organic–inorganic inhibitors

Eventually, research on improving the efficiency of anticorrosion compounds was continued resulting the fourth generation of corrosion inhibitors. Research has shown that an organic inhibitor next to an inorganic inhibitor has a synergistic effect and reduces the rate of corrosion in saline and neutral environments. It has also been reported that the direct use of organic and inorganic inhibitors together with the organic coatings compared to inorganic inhibitors can cause problems in the curing process and reduce the cross-linking density. To solve the problems caused by the direct use of organic and/ or inorganic inhibitors, organic–inorganic hybrids were synthesized. Hybrid nanostructures with low solubility and minimal effect in the curing process were proposed as the fourth generation of anticorrosion pigments, which had better active protection properties than the previous generations due to the synergism effect of the organic and inorganic parts. Alongside the coating phase, numerous studies are conducted in solution phase for corrosion control of the harsh media. As formerly explained, pristine organic materials are not substantial candidates for corrosion prevention and for tuning the inhibition impact organic–inorganic inhibitors are applied as alternatives.

With chelation of organic side with inorganic branch an inhibitor with new characteristics would appear (Figure 10.1). Coverage of both anodic and cathodic branch of the immersed sample is the first option. Actually, the organic compounds tend to chelate with anodic branches and the cathodic sides will remain free. It is a great chance for the corrosive media to destruct sample from these areas. However, in the presence of inorganic material in the hybrid organic–inorganic inhibitor, cathodic sides will cover by inorganic compounds which guarantee adequate protection. The other advantage of hybrid inhibitors compared to inorganic inhibitors is the formation of thicker film with higher density on the metal surface due to the larger size structure and lower solubility of organic molecules [2, 4, 7, 10].

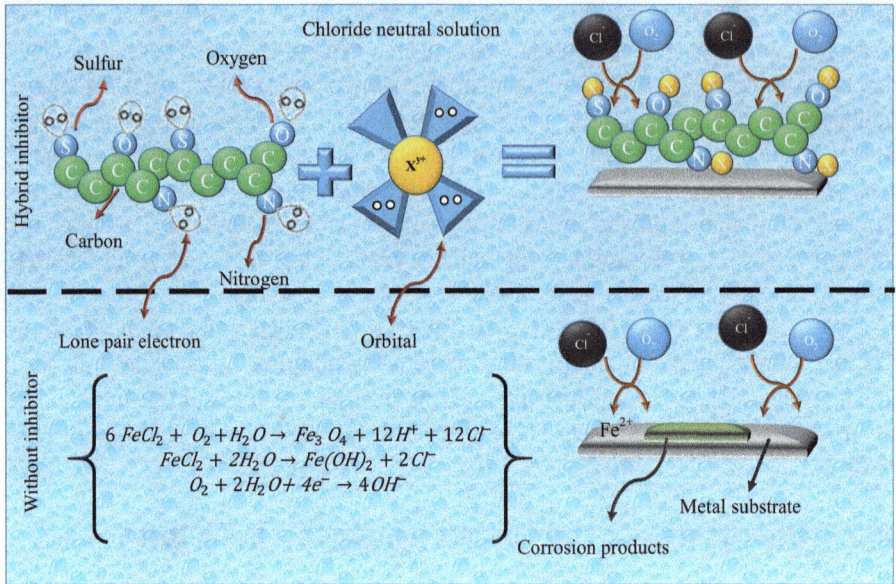

Figure 10.1: The schematic view of hybrid complex production and their protection mechanism.

10.1.4 Time to enter in nano era

Although organic–inorganic nanostructure hybrids indicated excellent inhibition effect in the solution media, there were some challenges in applying them into organic coatings and harsh solutions.

With no doubt, application of nanomaterials is a milestone in corrosion science. With excellent functions and tremendous abilities, researchers tried to bridge the traditional inhibitors with the nanomaterials to generate potential anticorrosion materials. Up to now, different nanomaterials including graphene oxide, carbon nanotube, MXenes, LDHs and h.BN are utilized for this purpose. Up to now, may efforts have been made to combine fourth generation of inhibitors (means organic–inorganic) with the available nanomaterials to design novel smart protective agents. However, since they had weak physical and noncovalent bonds with nanocarriers, desorption of active anticorrosion agents might occur during the coating curing process which had undesirable impression on barrier protection of coatings and the protection index in solution phase. Also, the inhibition performance of the nanostructures is limited owing to blinding the functional parts of nanocontainers upon adsorption of the organic–inorganic agents.

As a result, researchers used a new class of nanostructures; compounds with organic–inorganic structure have these properties. First, they should be able to create strong interactions such as coordination or covalent bonds with nanocontainers and can release inhibitor agents well. Second, they can solve the problem of the limitation of nanocarriers active sites to absorb inhibitors by growing on the carriers. So researchers apply metal-organic frameworks (MOFs) to overcome the challenges. They construct organic ligands and metal/metal oxide clusters that are chelated together by coordination connections (Figure 10.2). Also, they are slightly soluble in aqueous solutions so can release organic and inorganic inhibitive ions. The liberated metal cations react with OH^- anions on the cathodic sites and the organic parts absorb on the anodic area. On the other hand, MOFs also have the ability to grow on nanocontainers during synthesis process and are not limited to the functional parts which are placed on the nanocarrier surfaces [11, 12].

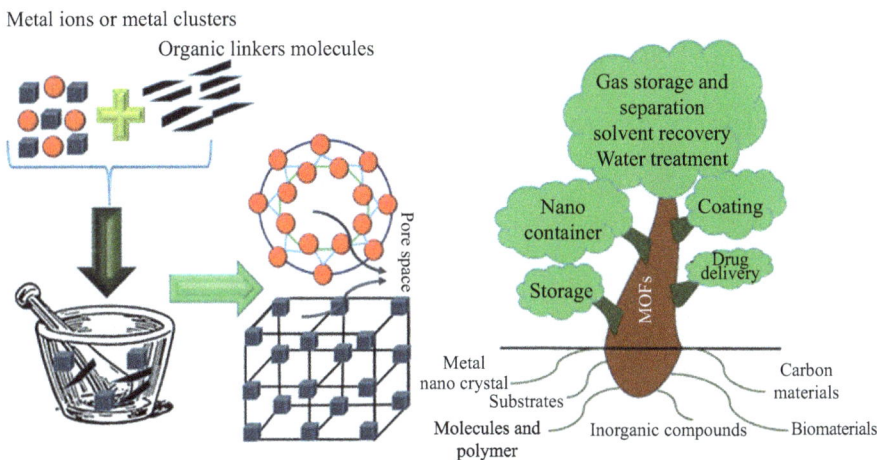

Figure 10.2: The schematic view of MOF productions, structures, classification and most utilization.

10.2.1 Application of Zn alongside organic compounds

The zinc is a metallic element and located as first element in group 12 of periodic table of elements with atomic number 30. The zinc element is frangible and crystalline at common temperature and is malleable at 110 up to 150 °C with the melting point and boiling point of Zn at 420 and 906 °C, respectively. In another side, the zinc ions are common from the positive charge cations with +2 charge number, but it rarely forms with +1 charge number. The Zn compound with +1 charge number needs some ligands to make this oxidation number stable. The good chemical properties of Zn can be resulted from +2 charge number cations.

In addition, the reactivity of zinc is good and it can easily be reacted by acidic or alkaline media. Due to +2 charge number of Zn, it is a great reducing agent and can easily form ionic bonding. The chemical properties of Zn are Ni and Cu (the end of transition metals first row) and the ionic radius of Zn is near Mg.

Between different applications of Zn (such as production of die-casting, electrical and hardware industries), it can play an important role in corrosion prevention of metals/alloys. As for good corrosion prevention of Zn and Zn compounds, it is more than a half of century that the Zn and Zn compounds are used in coatings for corrosion suppression all over the world [13–16]. With great smart utilization, researches displayed that the coatings containing the Zn particles can provide the cathodic protection as impediment protection. Zinc oxide (ZnO) is one common inorganic compound of zinc which is usually used as an additive in various materials and applications such as paints, coatings, medicine and corrosion preventions [17–20]. Nowadays, ZnO is widely applied as additive for paints for corrosion prevention of different metals in long periods [21–23].

According to unique properties of Zn salts (zinc nitrate [24], zinc acetylacetonate [25], zinc acetate [26], etc.), the utilization of zinc salts is highly appreciated as potential inhibitors in diverse studies. The main and foremost reason of switching from ZnO toward Zn salts is superior solubility of zinc salts against ZnO in water aqueous environments. Up to now, various researches illustrated the effect of zinc salts on corrosion prevention of metals. However, respect with the great solubility, numerous studies reflected that Zn salts are not reliable sources with high-indexed inhibition proficiency [27–29]. This is the main deriving force for researchers to open new perspective of designing inhibitive Zn-based materials.

The most straightforward approach is utilization of organic compounds besides Zn salts. Owing to the presence of heteroatoms in molecular structure of organic inhibitors, good functionality of these inhibitors is expected; but in some media, the inhibition efficiency of organic inhibitors is not acceptable [28, 30]. Simultaneously, due to good properties and function of zinc salts on cathodic protection, the organic inhibitors can combine zinc salts to maintain assembled nanocomposites with excellent mixed-type anticorrosion properties. According to the exhibition of heteroatoms in molecular structure of organic inhibitors and the charge of zinc, they can conversely react and by creation of ionic bonds, π bonds or other noncovalent reactions a unique assembled super molecule with excellent anticorrosion activity can be made.

One of the important inhibitors is called extracts, which are constructed from renewable parts of plants including seeds, leaves and roots. Due to their accessibility and environmental friendliness, the utilization of these kinds of inhibitors is widespread recently. Despite great inhibition efficiency of such resources in acidic media [31–33], they behave insufficiently in natural chloride media. This weakness is enough for researchers to look for a simple method to tune the performance of such inhibitors. Nowadays, various researches clarified that this idea is completely practical to maintain extremely acceptable performance in the mentioned neutral environments.

As an example, we can point to Bahlakeh et al. [34] who research on the effect of plant leaf extract and zinc synergism on carbon steel in chloride aqueous media. The results of their study indicated that 200:200 ppm of nettle leaf extract:Zn can mitigate the carbon steel corrosion impressively. In addition, in another study, Sanaei et al. [26] surveyed the influence of zinc acetate and *Cichorium intybus* leaf extracts on the mild steel corrosion in saline media. The outcomes displayed that the produced hybrid inhibitor can control mild steel corrosion in saline environment by mixed-type inhibition mechanism (both anodic and cathodic reactions were repressed). In another research, Salehi et al. [35] studied the inhibition effect of zinc acetate and *Urtica dioica* synergism on the mild steel corrosion in the neutral chloride media. The outcomes of high-resolution images (FE-SEM) indicated the production of protective film over the mild steel substrate. In addition, the electrochemical measurements demonstrate the mitigation of corrosion after introducing the mixture of zinc acetate and *Urtica dioica* into neutral chloride media. Similar works were reported by Majd et al. [36] (Persian *Echium amoenum*:zinc nitrate), Mofidabadi et al. [37] (*Sinapis arvenisis*: Zn nitrate) and Loto et al. (*Rosmarinus officinalis*:zinc oxide) [38], demonstrating that the combination of zinc salts has no contraction over the performance of extracts' impact and positively promote their performance in neutral chloride media.

There are various synthetic compounds caring double bonds or different heteroatoms, which are good for the inhibition but the time of synthesis process, high cost and the toxicity of these inhibitors were important factors which means application of high concentration of such resources can directly threaten the industries processing affordability alongside the environmental standards. Such awareness compelled the researchers to look for reliable alternate methods. With no doubt, a combination of low concentration of synthetic materials with inorganic compounds is a wise strategy for claiming excellent protection performance.

In one research, Zhang et al. [39] studied the synergistic effect of Zn^{2+} by polyaspartic acid, sodium gluconate and polyamino polyether methylenephosphonate on steel corrosion in soft water. The outcomes of this study indicated the mixed-type inhibition of inhibitors combination. In addition Baby and Manjula [40] surveyed the effect of $Zn(SO_4)$ and dimethyl pyrimidine on carbon steel corrosion in neutral aqueous. The results shown by introducing 80 ppm of Zn^{2+} the inhibition efficiency was achieved to 88.4%, which displayed the synergistic effect of Zn^{2+} and dimethyl pyrimidine. Moreover, the results of EL-Lateef and Alnajjar [41] on the enhancement of polytoluidine protection capacity for carbon steel by the use of zinc or lanthanum in acidic media displayed that by adding zinc or lanthanum the corrosion capacity of carbon steel was enhanced up to 96.8% or 98.9%, respectively. Mohamed et al. [42] investigated the synergistic effect of Zn and SO (sodium octanoate) on the carbon steel corrosion in cooling water. The outcomes demonstrate higher inhibition impact of SO: Zn mixture for carbon steel corrosion.

The biopolymer inhibitors are another famous kind of inhibitors, which were utilized in recent years. The utilization of this kind of inhibitor is spreading in recent years due to good properties such as nontoxicity and good inhibition efficiency [43–45]. There are various kinds of natural polymers such as exudate gums [46], starch [47], dextrin [48] and alginates [44]. Unfortunately there are a few number of researchers who combined biopolymers with zinc salts. Claims from these studies revealed the effect of zinc on the corrosion inhibition of biopolymers in harsh environments.

As an example, Kumar et al. [49] studied the effect of zinc on the corrosion inhibition of chitin and modified chitin on the mild steel corrosion in neutral chloride media. The results displayed the enhancement of corrosion inhibition of Zn/PCT up to 94%. In another research, Obasi [50] studied the effect of PVA and Zn particle mixture on the Al corrosion in mixed saline media. The outcomes displayed that the inhibition efficiency was reached to 86% by adding the ratio of 3:1 of PVA:Zn.

The ionic liquid-based inhibitor is another important organic corrosion inhibitor, which is consisted of organic nitrogenous compounds with a long hydrocarbons chain (the hydrocarbon chain had about 12 up to 18 carbon). The usual ionic liquids are produced by the combination of organic heterocyclic cations and organic or inorganic anions. As diverse scientific studies reflected, the inhibitor's impact is tied up with the presence of heteroatoms (O, N, S and P) in their structures, which are presented in heterocyclic compound. The heteroatoms have lone pair electron, which can be given to the free d-orbital of Fe or the metal vacant dπ orbitals interact by planar pπ orbitals of the aromatic rings. Also, adjustable hydrocarbon chain is another unique option which is available on ionic liquids. The presence of long hydrocarbon chain in molecular structure of ionic liquids can cover more area of metal substrate. According to the hydrophobic behavior of hydrocarbon chain the water molecules were rejected and the electrolyte cannot reach the metal substrate. Due to the inhibition ability of ionic liquids, they can be mixed by zinc salts to design a new class of hybrid inhibitors for limiting the neutral chloride media corrosion. However, to the best of our knowledge, this type of material is still under assessment and no results are provided. Herein, we suggest that this class of materials be considered for future corrosion orient works.

10.2.2 Cerium as powerful alternative

The second element in lanthanide groups with atomic number of 58, which is subset in rare earth materials, is called cerium. The oxidation number of cerium is often +3 but there is a stable form of +4 charge number. The cerium is metallic element with silvery white appearance. Among different applications of cerium (flat screen television, low energy light bubbles, coating of aluminum, etc.) the biodegradability and environmental-friendly properties of cerium are important factors which lead to the expansion of cerium use in different applications. According to the mentioned applications and properties of cerium, it can be used alongside of different inhibitors to improve the properties of inhibitors. There are different researches who investigated the effect of cerium and cerium alongside of organic inhibitors on corrosion prevention of metals.

As an example, Hu et al. [51] studied the influence of Ce on corrosion prevention of Al–Zn–Mg alloy. The results of their studies were displayed by introducing the excess amount of Ce, and the corrosion prevention and mechanical properties of mentioned alloy were reduced; but by adding 0.04 wt% of Ce the corrosion was mitigated significantly. Also in another research, Rodic and Milosev [52] investigated the effect of cerium(III) and cerium(IV) salts on corrosion prevention of Al and Al alloys in chloride media. The results indicated that the cerium(IV) does not inhibit as an inhibitor but the cerium(III) salts can effect on corrosion mitigation of Al and its alloys noticeably. However, the outcomes illustrated that the effectiveness of inhibitor depends on the types of substrate, anions and the salt concentrations.

Alike Zn salts, diverse researchers tried to promote organic compounds and inorganic Ce salt by making a synergistic effect between them. As formerly explained, green corrosion inhibitors are the most recommended materials among all types of inhibitors. With deep glance, we can realize that the extracts are the most applied green inhibitor owing to great characteristics. As a reason, the combination of Ce with plant extracts has been placed in the scientific line of many researchers.

For instance, Ramezanzadeh et al. [53] studied the synergistic effect of green nettle leaf extract and cerium nitrate on corrosion prevention of mild steel in 3.5% chloride solution. The result showed the fact that the concentration of inhibitor is an important factor which can affect the anticorrosion efficiency of inhibitors. In their study, the results shown by adding 400:400 ppm of nettle leaf extract:Ce, the inhibition efficiency was achieved to highest amount (about 95%). However, the outcomes demonstrated the mixed-type inhibition (both anodic and cathodic protection) of nettle leaf extract:Ce combination. In one research Dehghani et al. [54] investigated the synergistic effect of Ce and *Brassica hirta* extract on mild steel corrosion in saline solution. The results displayed that the optimum concentration of inhibitors mixture was 200:600 ppm and the maximum inhibition efficiency was about 92%. In addition in another study the same research team assessed the effect of quercetin and Ce mixture on the steel corrosion in neutral saline solution [55]. However the outcomes of their research showed the mixed-type inhibition mechanism of quercetin:Ce mixture and the maximum inhibition efficiency was reached to 96%.

Alongside extracts, diverse researchers preferred to analyze some synthetic materials and other types of green inhibitors as the organic side of hybrid material.

Ivušić et al. [56] surveyed the synergistic effect of cerium chloride and sodium gluconate on the carbon steel corrosion in sea water. The outcomes of their research displayed that the maximum inhibitions efficiency was achieved after the combination of cerium chloride and sodium gluconate (94.98%). Also Liu et al. [57] surveyed the synergistic effect of cerium nitrate:sodium dodecylbenzensulfonate on Al alloys in 3% saline solution. After using each inhibitors, the outcomes were displayed and the inhibition efficiency was limited; but by the combination of these inhibitors, the inhibition effectiveness was reached to 87.8%. Also the potentiodynamic results illustrated that by introducing the inhibitors mixture both anodic and cathodic reactions were retarded. Zhu et al. [58] in another research surveyed the effect of cerium(III) and glutamic acid combination on the Al alloy corrosion in neutral saline solution. The results of electrochemical impedance spectroscopy indicated the fact that the maximum inhibition efficiency was achieved to 85.4%. In addition, Dehghani et al. [59] studied the influence of methylphosphonic acid and trivalent cerium ions on mild steel corrosion in saline media. The outcomes of their study showed the effect of inhibitors concentration on the inhibition function. Therefore, the results illustrated by adding 400:400 ppm of methylphosphonic acid and trivalent cerium after 96 h immersion the inhibition effectiveness were about 93%. In addition, the Tafel analysis indicated the mixed-type inhibition of methylphosphonic acid and trivalent cerium on corrosion prevention of mild steel. There are more researches that investigated the effect of Ce (III) or (IV) on the inhibition effectiveness of inhibitors such as Dehghani (quercetin: Ce) [55], Li (anionic surfactant:Ce) [60] and Aramaki (sodium silicate:Ce) [61] representing the positive effect of cerium on the enhancement of inhibitor effectiveness.

10.2.3 Other materials

Due to good properties of inhibitors such as low cost and biodegradability, the utilization of hybrid inhibitor (mixture) is enhanced more and more nowadays. Therefore, the utilization of other rare earth materials and metal-based salts alongside of organic inhibitors in neutral chloride solution become a hotspot for verity of researchers.

For example, in one research, Dehghani et al. [62] surveyed the synergistic effect of europium(III) and quercetin on the mild steel corrosion in 3.5% saline solution. The outcomes displayed that the maximum inhibition efficiency was reached to 98%. In addition, the results indicated the effectiveness of inhibitors concentration on corrosion prevention, which was shown to be the best functionality of 200:600 ppm europium: quercetin. In another research, Usman et al. [63] surveyed the inhibition performance of KI and tannic acid on the steel corrosion into carbon dioxide saturated saline solution (3.5%). The electrochemical values demonstrated that by mixing TA and KI the maximum inhibition effectiveness was reached to 90% after 24 h immersion. The results of this study showed the effect of inhibitors concentration and the immersion time on the inhibition efficiency of inhibitors. Riazaty et al. [64] studied the inhibition effect of samarium:benzimidazole on the mild steel corrosion in saline media. The outcomes of electrochemical analysis and polarization test indicated the 92% inhibition efficiency with mixed-type inhibition respectively. In addition they showed the influence of inhibitors concentration on the corrosion mitigation (optimum concentration was 250:750 of Sm:BI). In addition Dehghani et al. [65] showed that by adding the mixture of Eu and benzimidazole on the corrosion prevention of steel in 3.5% neutral saline the solution was improved. The electrochemical results demonstrated that the inhibitor concentration is one of the most important factors that can effect on the steel corrosion mitigation. The outcomes affirming that by adding 200:600 ppm of benzimidazole:Eu, the maximum inhibition yield was attained to 98%. Volaric et al. [66] surveyed the effect of cerium and lanthanum combination on the corrosion mitigation of Al alloys in chloride media. The electrochemical analysis showed the synergistic effect of cerium: lanthanum mixture, which indicated the better inhibition impact of rare earth mixture. In addition, the EDS analysis confirmed the presence of both rare earth cations.

10.2.4 Metal-organic frameworks

The weakness points of conventional corrosion inhibitors convinced researchers to open new gates toward mixed and hybrid inhibitors. However, with detailed insights over such inhibitors, more and more reasons have been appeared for scientists who are responsible for walking through more responsive inhibitors with acceptable performance. Right now by the utilization of MOFs, with coordinated structures, the weakness points of the previous classes of inhibitors are resolved [67]. The MOFs with the control-release capability can supply acceptable portion of inhibitors into corrosive media over the time, concluding the remarkable corrosion retardation [68]. Other positive characteristics of MOF utilization against hybrid inhibitors are their pH sensitivity; that is, they are quite responsive against corrosive reactions [69]. Also, due to the exhibition of both metallic clusters and the organic ligands in MOF structures, they potentially govern metals against corrosion via mixed-type inhibition mechanisms [70]. Therefore, the demand of MOF application in corrosion has been explosively spread out alongside other applications [71].

With a simple look over MOF clusters can realize that different metallic elements such as Zn, Zr and Ce can be employed [72–74]. For instance, one functional class of MOFs is zeolite imidazole framework-8 (ZIF-8) which is applied in the coating to protect the metal substrate. They are porous hybrid materials with similar structure to zeolites and are constructed upon four connected nets of tetrahedral units [11, 75]. The dissolution of MOFs (ZIF-8 as an example) in saline media was resulted from the liberation of metallic cations and organic molecules [76, 77]. The organic components (2-methylimidazole in ZIF-8) adsorb over the anodic sites by the donation of lone pair electron. In another side the released metallic cations from MOF structure (Zn) react by hydroxide anions and constitute a protective layer over the cathodic area [11]. With the unchanged protection mechanism, various MOFs including ZIFs, MILs and UiOs can aid corrosion suppression in harsh environments to induce excellent active corrosion protection properties [11, 12, 78, 79].

However, alongside great anticorrosion impact of MOFs, their porous structures gave them an exceptional opportunity for inhibitors encapsulation in their galleries [80]. As a result, after the MOFs decomposition against environmental stimuli, the loaded inhibitors will release as a function of time that guarantees superior protection (Figure 10.3). There are different researches that illustrated the good effect of MOF-loaded inhibitors on corrosion prevention of steel or other alloys in corrosive media [69, 81–88].

10.2.5 Suggestions and future perspectives

Due to good functionality of MOFs for corrosion aspects, there are various proposals about the tuning the effect of MOFs as far more possible. As a simple perspective,

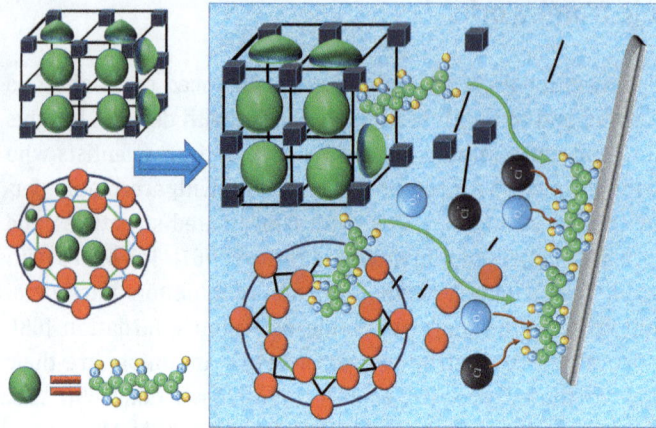

Figure 10.3: The schematic view of MOF decomposition and liberation of corrosion inhibitors over the time.

varying metallic cations such as Eu, La, Pr and Sm probably can impose some remarkable effects on the MOFs functionality. Alongside, the utilization of MOFs loaded with various organic inhibitors is a great suggestion for future research.

Apart from all, the nanosized structure of MOFs gives the opportunity to synthesize over lamellar structures, including LDHs, GO/graphene, MXenes and MWCNTs. With following this concept, more exceptional complicated nanohybrids can be diagnosed, which are surely practical for corrosion issues.

10.3 Conclusion

Due to the low cost and good inhibition efficiency of corrosion inhibitor, the utilization of corrosion inhibitor is slipped over designing more potential cases. Nowadays, due to global concern about the environmental hazards, driving forces are imposed over researchers to look for modern alternatives for application of traditional methods. To compensate such weakness point, researchers grabbed the utilization of hybrid inhibitor. However, despite enriching tremendous outcomes, the first class of hybrid is unable to convince all environmental/industrial standards. The inability in adding corrosion inhibitor in aggressive media can be another weakness point of the corrosion inhibitor usage. As a reason, with proposing sophisticated ideas strong steps were taken to develop the next generations of hybrids (means nanohybrids). Conclusively, emergence of MOFs significantly responded to researchers' demands and concerns. With excellent responsivity against pH and control-release capability, MOFs are nominated as the most advanced nano-hybrids with exceptional protection for long-lasting periods.

References

[1]   McCafferty, E., Introduction to Corrosion Science, Springer, New York, 2010.

[2]   Ralkhal, S., Ramezanzadeh, B., Shahrabi, T., Studying dual active/barrier and self-healing reinforcing effects of the Neodymium (III)-Benzimidazole hybrid complex in the epoxy coating/mild steel system, Journal of Alloys and Compounds, 2019, 790, 141–155.

[3]   Xie, Y., Liu, C., Liu, W., Liang, L., Wang, S., Zhang, F., et al., A novel approach to fabricate polyacrylate modified graphene oxide for improving the corrosion resistance of epoxy coatings, Colloids and Surfaces A: Physicochemical and Engineering Aspects, 2020, 593, 124627.

[4]   Umoren, S.A., Solomon, M.M., Synergistic corrosion inhibition effect of metal cations and mixtures of organic compounds: A review, Journal of Environmental Chemical Engineering, 2017, 5, 246–273.

[5]   Hao, Y., Liu, F., Han, E.-H., Anjum, S., Xu, G., The mechanism of inhibition by zinc phosphate in an epoxy coating, Corrosion Science, 2013, 69, 77–86.

[6]   Sanaei, Z., Ramezanzadeh, B., Shahrabi, T., Anti-corrosion performance of an epoxy ester coating filled with a new generation of hybrid green organic/inorganic inhibitive pigment; electrochemical and surface characterizations, Applied Surface Science, 2018, 454, 1–15.

[7]   Abrishami, S., Naderi, R., Ramezanzadeh, B., Fabrication and characterization of zinc acetylacetonate/*Urtica dioica* leaves extract complex as an effective organic/inorganic hybrid corrosion inhibitive pigment for mild steel protection in chloride solution, Applied Surface Science, 2018, 457, 487–496.

[8]   Mousavifard, S., Nouri, P.M., Attar, M., Ramezanzadeh, B., The effects of zinc aluminum phosphate (ZPA) and zinc aluminum polyphosphate (ZAPP) mixtures on corrosion inhibition performance of epoxy/polyamide coating, Journal of Industrial and Engineering Chemistry, 2013, 19, 1031–1039.

[9]   Hamadi, L., Mansouri, S., Oulmi, K., Kareche, A., The use of amino acids as corrosion inhibitors for metals: A review, Egyptian Journal of Petroleum, 2018, 27, 1157–1165.

[10] Ramezanzadeh, B., Ghasemi, E., Askari, F., Mahdavian, M., Synthesis and characterization of a new generation of inhibitive pigment based on zinc acetate/benzotriazole: Solution phase and coating phase studies, Dyes and Pigments, 2015, 122, 331–345.

[11] Ramezanzadeh, M., Ramezanzadeh, B., Mahdavian, M., Bahlakeh, G., Development of metal-organic framework (MOF) decorated graphene oxide nanoplatforms for anti-corrosion epoxy coatings, Carbon, 2020, 161, 231–251.

[12] Jiang, L., Dong, Y., Yuan, Y., Zhou, X., Liu, Y., Meng, X., Recent advances of metal–organic frameworks in corrosion protection: From synthesis to applications, Chemical Engineering Journal, 2022, 430, 132823.

[13] Shreepathi, S., Bajaj, P., Mallik, B., Electrochemical impedance spectroscopy investigations of epoxy zinc rich coatings: Role of Zn content on corrosion protection mechanism, Electrochimica Acta, 2010, 55, 5129–5134.

[14] Park, J.H., Yun, T.H., Kim, K.Y., Song, Y.K., Park, J.M., The improvement of anticorrosion properties of zinc-rich organic coating by incorporating surface-modified zinc particle, Progress in Organic Coatings, 2012, 74, 25–35.

[15] Meroufel, A., Touzain, S., EIS characterisation of new zinc-rich powder coatings, Progress in Organic Coatings, 2007, 59, 197–205.

[16] Pereira, D., Scantlebury, J., Ferreira, M., Almeida, M., The application of electrochemical measurements to the study and behaviour of zinc-rich coatings, Corrosion Science, 1990, 30, 1135–1147.

[17] Gu, X., Li, C., Yuan, S., Ma, M., Qiang, Y., Zhu, J., ZnO based heterojunctions and their application in environmental photocatalysis, Nanotechnology, 2016, 27, 402001.

[18] Oprea, O., Andronescu, E., Ficai, D., Ficai, A., Oktar F, N., Yetmez, M., ZnO applications and challenges, Current Organic Chemistry, 2014, 18, 192–203.

[19] Di Mauro, A., Fragala, M.E., Privitera, V., Impellizzeri, G., ZnO for application in photocatalysis: From thin films to nanostructures, Materials Science in Semiconductor Processing, 2017, 69, 44–51.

[20] Klingshirn, C., ZnO: From basics towards applications, Physica Status Solidi (B), 2007, 244, 3027–3073.

[21] Ramezanzadeh, B., Attar, M., Studying the effects of micro and nano sized ZnO particles on the corrosion resistance and deterioration behavior of an epoxy-polyamide coating on hot-dip galvanized steel, Progress in Organic Coatings, 2011, 71, 314–328.

[22] Ramezanzadeh, B., Attar, M., An evaluation of the corrosion resistance and adhesion properties of an epoxy-nanocomposite on a hot-dip galvanized steel (HDG) treated by different kinds of conversion coatings, Surface and Coatings Technology, 2011, 205, 4649–4657.

[23] Naing, T.H., Janudom, S., Mahathaninwong, N., Rachpech, V., Karrila, S., Morphology and corrosion behavior of ZnO passive films for galvanized steel applications: Effects of anodizing parameters, Surface Topography: Metrology and Properties, 2022, 10, 025005.

[24] Kasaeian, M., Ghasemi, E., Ramezanzadeh, B., Mahdavian, M., Bahlakeh, G., A combined experimental and electronic-structure quantum mechanics approach for studying the kinetics and adsorption characteristics of zinc nitrate hexahydrate corrosion inhibitor on the graphene oxide nanosheets, Applied Surface Science, 2018, 462, 963–979.

[25] Dehghani, A., Bahlakeh, G., Ramezanzadeh, B., Beta-cyclodextrin-zinc acetylacetonate (β-CD@ZnA) inclusion complex formation as a sustainable/smart nanocarrier of corrosion inhibitors for a water-based siliconized composite film: Integrated experimental analysis and fundamental computational electronic/atomic-scale simulation, Composites Part B: Engineering, 2020, 197, 108152.

[26] Sanaei, Z., Shahrabi, T., Ramezanzadeh, B., Synthesis and characterization of an effective green corrosion inhibitive hybrid pigment based on zinc acetate-*Cichorium intybus* L leaves extract (ZnA-CIL. L): Electrochemical investigations on the synergistic corrosion inhibition of mild steel in aqueous chloride solutions, Dyes and Pigments, 2017, 139, 218–232.

[27] Mofidabadi, A.H.J., Dehghani, A., Bahlakeh, G., Ramezanzadeh, B., Combined clove extract bio-molecules and zinc(II) ion synergistic effects in steel corrosion mitigation in saline solution: Electronic (DFT) modeling, atomic/molecular (MC/MD) simulations, and corrosion measurement, Biomass Conversion and Biorefinery, 2022.

[28] A Hossein Jafari, M., Dehghani, A., Ramezanzadeh, B., *Sinapis arvensis* (Mustard) extract derived bio-molecules linked Zinc-II ions; Integrated electrochemical & surface investigations, Journal of Molecular Liquids, 2022, 346, 117085.

[29] Mofidabadi, A.H.J., Dehghani, A., Ramezanzadeh, B., Steel-alloy surface protection against saline attacks via the development of Zn(II)-metal-organic networks using Lemon verbena leaves extract (LVLE); Integrated surface/electrochemical explorations, Colloids and Surfaces A: Physicochemical and Engineering Aspects, 2021, 630, 127561.

[30] Mofidabadi, A.H.J., Dehghani, A., Ramezanzadeh, B., Investigating the effectiveness of Watermelon extract-zinc ions for steel alloy corrosion mitigation in sodium chloride solution, Journal of Molecular Liquids, 2022, 346, 117086.

[31] Dehghani, A., Ramezanzadeh, B., Rosemary extract inhibitive behavior against mild steel corrosion in tempered 1 M HCl media, Industrial Crops and Products, 2023, 193, 116183.

[32] Nazari, A., Ramezanzadeh, B., Guo, L., Dehghani, A., Application of green active bio-molecules from the aquatic extract of Mint leaves for steel corrosion control in hydrochloric acid (1M) solution: Surface, electrochemical, and theoretical explorations, Colloids and Surfaces A: Physicochemical and Engineering Aspects, 2023, 656, 130540.

[33] Ghahremani, P., Mostafatabr, A.H., Dehghani, A., Bahlakeh, G., Ramezanzadeh, B., Apple pomace extract: A potent renewable source of active biomolecules for suppressing mild steel aggression in aquatic solution, Biomass Conversion and Biorefinery, 2022.

[34] Bahlakeh, G., Ramezanzadeh, M., Ramezanzadeh, B., Experimental and theoretical studies of the synergistic inhibition effects between the plant leaves extract (PLE) and zinc salt (ZS) in corrosion control of carbon steel in chloride solution, Journal of Molecular Liquids, 2017, 248, 854–870.

[35] Salehi, E., Naderi, R., Ramezanzadeh, B., Synthesis and characterization of an effective organic/inorganic hybrid green corrosion inhibitive complex based on zinc acetate/Urtica Dioica, Applied Surface Science, 2017, 396, 1499–1514.

[36] Bahlakeh, G., Dehghani, A., Ramezanzadeh, B., Ramezanzadeh, M., A green complex film based on the extract of Persian Echium amoenum and zinc nitrate for mild steel protection in saline solution; Electrochemical and surface explorations besides dynamic simulation, Journal of Molecular Liquids, 2019, 291, 111281.

[37] Mofidabadi, A.H.J., Dehghani, A., Ramezanzadeh, B., *Sinapis arvensis* (Mustard) extract derived biomolecules linked Zinc-II ions; Integrated electrochemical & surface investigations, Journal of Molecular Liquids, 2022, 346, 117085.

[38] Loto, R.T., Surface coverage and corrosion inhibition effect of *Rosmarinus officinalis* and zinc oxide on the electrochemical performance of low carbon steel in dilute acid solutions, Results in Physics, 2018, 8, 172–179.

[39] Zhang, B., He, C., Wang, C., Sun, P., Li, F., Lin, Y., Synergistic corrosion inhibition of environment-friendly inhibitors on the corrosion of carbon steel in soft water, Corrosion Science, 2015, 94, 6–20.

[40] Nirmala Baby, D.P., Dimethyl pyrimidine derivative–an environmentally friendly corrosion inhibitor for inhibition of corrosion of carbon steel in neutral aqueous environment, International Journal of Engineering Science, 2016, 6849.

[41] Abd El-Lateef, H.M., Alnajjar, A.O., Enhanced the protection capacity of poly (o-toluidine) by synergism with zinc or lanthanum additives at C-steel/HCl interface: A combined DFT, molecular dynamic simulations and experimental methods, Journal of Molecular Liquids, 2020, 303, 112641.

[42] Mohamed, K., Ibrahim, O., El-Bedawy, M., Ali, A., Synergistic effect of different Zn salts with sodium octanoate on the corrosion inhibition of carbon steel in cooling water, Journal of Radiation Research and Applied Sciences, 2020, 13, 276–287.

[43] Shahini, M., Ramezanzadeh, B., Mohammadloo, H.E., Recent advances in biopolymers/carbohydrate polymers as effective corrosion inhibitive macro-molecules: A review study from experimental and theoretical views, Journal of Molecular Liquids, 2021, 325, 115110.

[44] Jmiai, A., El Ibrahimi, B., Tara, A., El Issami, S., Jbara, O., Bazzi, L., Alginate biopolymer as green corrosion inhibitor for copper in 1 M hydrochloric acid: Experimental and theoretical approaches, Journal of Molecular Structure, 2018, 1157, 408–417.

[45] Obot, I., Onyeachu, I.B., Kumar, A.M., Sodium alginate: A promising biopolymer for corrosion protection of API X60 high strength carbon steel in saline medium, Carbohydrate Polymers, 2017, 178, 200–208.

[46] Mobin, M., Rizvi, M., Olasunkanmi, L.O., Ebenso, E.E., Biopolymer from Tragacanth gum as a green corrosion inhibitor for carbon steel in 1 M HCl solution, ACS Omega, 2017, 2, 3997–4008.

[47] Sushmitha, Y., Rao, P., Material conservation and surface coating enhancement with starch-pectin biopolymer blend: A way towards green, Surfaces and Interfaces, 2019, 16, 67–75.

[48] Biswas, A., Das, D., Lgaz, H., Pal, S., Nair, U.G., Biopolymer dextrin and poly (vinyl acetate) based graft copolymer as an efficient corrosion inhibitor for mild steel in hydrochloric acid: Electrochemical, surface morphological and theoretical studies, Journal of Molecular Liquids, 2019, 275, 867–878.

[49] Kumar, V., AR, B.V., Chemically modified biopolymer as an eco-friendly corrosion inhibitor for mild steel in a neutral chloride environment, New Journal of Chemistry, 2017, 41, 6278–6289.

[50] Obasi, I.B., Corrosion inhibition of Polyvinyl Alcohol (PVA), ZN nano particles for AA6061, Mixed Salt Solution.

[51] Hu, G., Zhu, C., Xu, D., Dong, P., Chen, K., Effect of cerium on microstructure, mechanical properties and corrosion properties of Al-Zn-Mg alloy, Journal of Rare Earths, 2021, 39, 208–216.

[52] Rodič, P., Milošev, I., Corrosion inhibition of pure aluminium and alloys AA2024-T3 and AA7075-T6 by cerium (III) and cerium (IV) salts, Journal of the Electrochemical Society, 2015, 163, C85.

[53] Ramezanzadeh, M., Sanaei, Z., Bahlakeh, G., Ramezanzadeh, B., Highly effective inhibition of mild steel corrosion in 3.5% NaCl solution by green Nettle leaves extract and synergistic effect of eco-friendly cerium nitrate additive: Experimental, MD simulation and QM investigations, Journal of Molecular Liquids, 2018, 256, 67–83.

[54] Dehghani, A., Mostafatabar, A.H., Ramezanzadeh, B., Synergistic anticorrosion effect of Brassica Hirta phytoconstituents and cerium ions on mild steel in saline media: Surface and electrochemical evaluations, Colloids and Surfaces A: Physicochemical and Engineering Aspects, 2023, 656, 130503.

[55] Dehghani, A., Bahlakeh, G., Ramezanzadeh, B., Hossein Mostafatabar, A., Ramezanzadeh, M., Estimating the synergistic corrosion inhibition potency of (2-(3, 4-)-3, 5, 7-trihydroxy-4H-chromen-4-one) and trivalent-cerium ions on mild steel in NaCl solution, Construction and Building Materials, 2020, 261, 119923.

[56] Ivušić, F., Lahodny-Šarc, O., Ćurković, H.O., Alar, V., Synergistic inhibition of carbon steel corrosion in seawater by cerium chloride and sodium gluconate, Corrosion Science, 2015, 98, 88–97.

[57] Liu, J., Wang, D., Gao, L., Zhang, D., Synergism between cerium nitrate and sodium dodecylbenzenesulfonate on corrosion of AA5052 aluminium alloy in 3 wt.% NaCl solution, Applied Surface Science, 2016, 389, 369–377.

[58] Zhu, C., Yang, H.X., Wang, Y.Z., Zhang, D.Q., Chen, Y., Gao, L.X., Synergistic effect between glutamic acid and rare earth cerium (III) as corrosion inhibitors on AA5052 aluminum alloy in neutral chloride medium, Ionics, 2019, 25, 1395–1406.

[59] Dehghani, A., Ramezanzadeh, B., Poshtiban, F., Bahlakeh, G., Construction of a highly-effective/sustainable corrosion protective composite nanofilm based on Aminotris (methylphosphonic acid) and trivalent cerium ions on mild steel against chloride solution, Construction and Building Materials, 2020, 261, 119838.

[60] Li, X., Deng, S., Fu, H., Mu, G., Synergistic inhibition effect of rare earth cerium (IV) ion and anionic surfactant on the corrosion of cold rolled steel in H2SO4 solution, Corrosion Science, 2008, 50, 2635–2645.

[61] Aramaki, K., Synergistic inhibition of zinc corrosion in 0.5 M NaCl by combination of cerium (III) chloride and sodium silicate, Corrosion Science, 2002, 44, 871–886.

[62] Dehghani, A., Mostafatabar, A.H., Bahlakeh, G., Ramezanzadeh, B., A detailed study on the synergistic corrosion inhibition impact of the Quercetin molecules and trivalent europium salt on mild steel; electrochemical/surface studies, DFT modeling, and MC/MD computer simulation, Journal of Molecular Liquids, 2020, 316, 113914.

[63] Usman, B.J., Umoren, S.A., Gasem, Z.M., Inhibition of API 5L X60 steel corrosion in CO2-saturated 3.5% NaCl solution by tannic acid and synergistic effect of KI additive, Journal of Molecular Liquids, 2017, 237, 146–156.

[64] Riazaty, P., Naderi, R., Ramezanzadeh, B., Synergistic corrosion inhibition effects of benzimidazole-samarium (III) molecules on the steel corrosion prevention in simulated seawater, Journal of Molecular Liquids, 2019, 296, 111801.

[65] Dehghani, A., Bahlakeh, G., Ramezanzadeh, B., Mofidabadi, A.H.J., Construction of a high-potency anti-corrosive metal-organic film based on europium (III)-benzimidazole: Theoretical and electrochemical investigations, Construction and Building Materials, 2021, 269, 121271.

[66] Volaric, B., Rodic, P., Milosev, I., Cerium and lanthanum salts used as individual and combined inhibitors for corrosion protection of AA7075-T6 in chloride solution, ECS Meeting Abstracts: IOP Publishing, 2015, 685.

[67] Zhang, X., Zhang, Y., Wang, Y., Gao, D., Zhao, H., Preparation and corrosion resistance of hydrophobic zeolitic imidazolate framework (ZIF-90) film@ Zn-Al alloy in NaCl solution, Progress in Organic Coatings, 2018, 115, 94–99.

[68] Dehghani, A., Sanaei, Z., Fedel, M., Ramezanzadeh, M., Mahdavian, M., Ramezanzade, B., Fabrication of an intelligent anti-corrosion silane film using a MoO42− loaded Micro/mesoporous ZIF67-MOF/multi-walled-CNT/APTES core-shell nano-container, Colloids and Surfaces A: Physicochemical and Engineering Aspects, 2023, 656, 130511.

[69] Alipanah, N., Dehghani, A., Abdolmaleki, M., Bahlakeh, G., Ramezanzadeh, B., Designing environmentally-friendly pH-responsive self-redox polyaniline grafted graphene oxide nano-platform decorated by zeolite imidazole ZIF-9 MOF for achieving smart functional epoxy-based anti-corrosion coating, Journal of Environmental Chemical Engineering, 2023, 11, 109048.

[70] Abdi, J., Izadi, M., Bozorg, M., Improvement of anti-corrosion performance of an epoxy coating using hybrid UiO-66-NH2/carbon nanotubes nanocomposite, Scientific Reports, 2022, 12, 1–14.

[71] Li, H., Qiang, Y., Zhao, W., Zhang, S., 2-Mercaptobenzimidazole-inbuilt metal-organic-frameworks modified graphene oxide towards intelligent and excellent anti-corrosion coating, Corrosion Science, 2021, 191, 109715.

[72] Abid, H.R., Tian, H., Ang, H.-M., Tade, M.O., Buckley, C.E., Wang, S., Nanosize Zr-metal organic framework (UiO-66) for hydrogen and carbon dioxide storage, Chemical Engineering Journal, 2012, 187, 415–420.

[73] Ren, B., Chen, Y., Li, Y., Li, W., Gao, S., Li, H., et al., Rational design of metallic anti-corrosion coatings based on zinc gluconate@ ZIF-8, Chemical Engineering Journal, 2020, 384, 123389.

[74] Fouda, -A.E.-A.S., Etaiw, S.E.-D.H., El-bendary, M.M., Maher, M.M., Metal-organic frameworks based on silver (I) and nitrogen donors as new corrosion inhibitors for copper in HCl solution, Journal of Molecular Liquids, 2016, 213, 228–234.

[75] Lee, Y.-R., Jang, M.-S., Cho, H.-Y., Kwon, H.-J., Kim, S., Ahn, W.-S., ZIF-8: A comparison of synthesis methods, Chemical Engineering Journal, 2015, 271, 276–280.

[76] Zhang, M., Liu, Y., Enhancing the anti-corrosion performance of ZIF-8-based coatings via microstructural optimization, New Journal of Chemistry, 2020, 44, 2941–2946.

[77] Mohammadkhani, R., Ramezanzadeh, M., Fedel, M., Ramezanzadeh, B., Mahdavian, M., PO43-Loaded ZIF-8-type Metal–Organic Framework-Decorated Multiwalled Carbon Nanotube Synthesis and Application in Silane Coatings for Achieving a Smart Corrosion Protection Performance, Industrial & Engineering Chemistry Research, 2022, 61, 11747–11765.

[78] Ramezanzadeh, M., Ramezanzadeh, B., Bahlakeh, G., Tati, A., Mahdavian, M., Development of an active/barrier bi-functional anti-corrosion system based on the epoxy nanocomposite loaded with highly-coordinated functionalized zirconium-based nanoporous metal-organic framework (Zr-MOF), Chemical Engineering Journal, 2021, 408, 127361.

[79] Tarzanagh, Y.J., Seifzadeh, D., Rajabalizadeh, Z., Habibi-Yangjeh, A., Khodayari, A., Sohrabnezhad, S., Sol-gel/MOF nanocomposite for effective protection of 2024 aluminum alloy against corrosion, Surface and Coatings Technology, 2019, 380, 125038.

[80] Xiong, L., Liu, J., Yu, M., Li, S., Improving the corrosion protection properties of PVB coating by using salicylaldehyde@ ZIF-8/graphene oxide two-dimensional nanocomposites, Corrosion Science, 2019, 146, 70–79.

[81] Choudhary, P., Kumari Ola, S., Chopra, I., Dhayal, V., Singh Shekhawat, D., Metal–organic framework (MOF)/graphene–oxide (GO) nanocomposites materials: A potential formulation for anti-corrosive coatings- a review, Materials Today: Proceedings, 2022.

[82] Jiang, S., Zhang, Z., Wang, D., Wen, Y., Peng, N., Shang, W., Zif-8-based micro-arc oxidation composite coatings enhanced the corrosion resistance and superhydrophobicity of a Mg alloy, Journal of Magnesium and Alloys, 2021.

[83] Tian, H., Li, W., Liu, A., Gao, X., Han, P., Ding, R., et al., Controlled delivery of multi-substituted triazole by metal-organic framework for efficient inhibition of mild steel corrosion in neutral chloride solution, Corrosion Science, 2018, 131, 1–16.

[84] Alipanah, N., Yari, H., Mahdavian, M., Ramezanzadeh, B., Bahlakeh, G., MIL-88A (Fe) filler with duplicate corrosion inhibitive/barrier effect for epoxy coatings: Electrochemical, molecular simulation, and cathodic delamination studies, Journal of Industrial and Engineering Chemistry, 2021, 97, 200–215.

[85] Nejad, S.A.T., Ramezanzadeh, M., Alibakhshi, E., Ramezanzadeh, B., Olivier, M.-G., Mahdavian, M., Fabrication of 8-hydroxyquinoline loaded in an aluminum-based metal-organic framework for strengthening anti-corrosion behavior of silane primer coating, Progress in Organic Coatings, 2023, 174, 107280.

[86] Lashgari, S.M., Yari, H., Mahdavian, M., Ramezanzadeh, B., Bahlakeh, G., Ramezanzadeh, M., Application of nanoporous cobalt-based ZIF-67 metal-organic framework (MOF) for construction of an epoxy-composite coating with superior anti-corrosion properties, Corrosion Science, 2021, 178, 109099.

[87] Cao, K., Yu, Z., Yin, D., Preparation of Ce-MOF@ TEOS to enhance the anti-corrosion properties of epoxy coatings, Progress in Organic Coatings, 2019, 135, 613–621.

[88] Keshmiri, N., Najmi, P., Ramezanzadeh, M., Ramezanzadeh, B., Designing an eco-friendly lanthanide-based metal organic framework (MOF) assembled graphene-oxide with superior active anti-corrosion performance in epoxy composite, Journal of Cleaner Production, 2021, 319, 128732.

Manash Protim Mudoi*, Rhythm Katyal, Khushi Bhatt, Vidushi Singh,
Asmita Choudhary, Sanskriti Gupta

Chapter 11
Ceramic nanomaterials in corrosion prevention

Abstract: Corrosion can be effectively mitigated by materials' surface modification without changing the bulk properties. Nanotechnology, with the development of nanocoatings over the material surface, is providing a better solution to the corrosion problem. The operational lifetime and the performance of materials enhanced with such protective coatings. It can provide other advantages like improved hardness, adhesion strength and tribological properties. The possibility of synthesizing the coatings either in thinner or thicker thickness allows better efficiency, lower operating maintenance costs, reduced carbon footprints and better fuel economy. This chapter discusses the different types of ceramic nanocoatings, routes of synthesis, properties and corrosion performance. The challenges are also highlighted.

Keywords: Ceramics, nanocoatings, corrosion, nanotechnology, surface modification

11.1 Introduction

Materials that are formed and generated with at least one dimension less than 100 nm in particle size are referred to as nanomaterials (NMs) [1]. In addition to being incredibly small, NMs also possess finest characteristics that set them apart from their bulk counterparts, making them strong contenders for a variety of applications [2]. These characteristics include high chemical reactivity, high surface area and high mechanical and thermal stability. Metals are effectively protected against corrosion by coatings, which are widely utilized. Howbeit, the longstanding corrosion resistance effect of the coating material continuously declines due to the old coating materials' poor resistance to the permeation of solutions with the ability to corrode at the coating–metal contact.

*Corresponding author: Manash Protim Mudoi, Department of Chemical Engineering, Indian Institute of Technology, Roorkee 247667, Uttarakhand, India; Department of Chemical Engineering, University of Petroleum and Energy Studies, Dehradun 248007, Uttarakhand, India,
e-mail: mp_mudoi@ch.iitr.ac.in, mpmudoi@ddn.upes.ac.in
Rhythm Katyal, Khushi Bhatt, Vidushi Singh, Asmita Choudhary, Sanskriti Gupta, Department of Chemical Engineering, University of Petroleum and Energy Studies, Dehradun 248007, Uttarakhand, India

https://doi.org/10.1515/9783111071756-011

Recently, coatings were combined with NMs to greatly enhance their mechanical, optical and chemical properties in order to overcome this limitation. NM-based coatings are either composed of nanoscale components or are constructed of thin layers with a thickness of less than 100 nm. Because of numerous advantages like the hardness of the surface, the strength of the adhesive and the resistance to corrosion over time and at high temperatures, NM-based coatings can be successfully used to reduce the undesirable effects of a corrosive medium. In addition, NM-based coatings can be employed efficiently with smoother thickness, allowing for greater adaptability in the construction of metal equipment and a decrease in process costs. An overview of current developments in the corrosive applications of ceramic-based NMs is given in this chapter. The characteristics of NMs are covered first, and then the various NM varieties are briefly discussed. Last but not least, corrosion applications of NMs are also discussed.

11.2 Behavior of nanomaterials

11.2.1 Adsorption behavior

The large adsorption of NMs is another fascinating characteristic that has drawn much interest from academics in recent years due to its potential environmental implications [3]. NMs enable physical adsorption processes on chemically inert surfaces. Furthermore, their huge surface areas are greater than those of conventional adsorbents like activated carbon. Activated carbon and various NMs, including carbon nanotubes (CNTs), are fundamentally distinct from one another. For instance, their structure is more uniform and well-defined at the nanoscale. While the numerous clearly specified adsorption sites that the adsorbed compounds have access to on CNTs can be dealt with directly, pore diameter and adsorption energy are two factors that must be taken into account while researching the adsorption on activated carbon.

11.2.2 Thermal stability

The distinct advantages of NMs, including their electrical [4, 5], optical [6, 7] and magnetic properties [8], have been well studied in the literature. However, the thermal characteristics of NMs have only recently received substantial attention. This is particularly true because it is difficult to assess and screen thermal transport in nanoscale dimensions. An effective method of measuring the thermal transport of NMs with high precision was proposed for this purpose, and this opens up a promising path for evaluating the thermal characteristics of NMs.

11.2.3 Mechanical stability

The tremendous complexity of NMs [9, 10] is one of the most striking features. Very hard nanocomposite materials can be made by successfully depositing borides, nitrides and carbides using plasma-induced chemical vapor deposition (CVD) and physical vapor deposition techniques. The stiffness property of the developed nanocomposite material performs significantly better than that offered by the bulk mixing rule in the properly prepared binary systems. The discovery of CNTs stimulated extensive research [11] because of their exceptional mechanical properties. The strength of carbon-based nanofibers is increased by graphitization along the fiber axis. CNTs are comprised of seamless cylindrical graphene nanolayers and are one of various forms of carbon fibers; hence, they are the best in terms of mechanical properties. Rapid development of mechanical features in NMs is essential when we mix NMs such as CNTs, nanocrystals, nanoparticles and nanowires. When analyzing the mechanical properties of NMs in depth, a few factors that should be carefully taken into account include the surface structure, functionalization, porosity, synthesis techniques and chemical changes. These factors have a big effect on the mechanical properties of NMs.

11.2.4 Electrical behavior

The electronic properties of NMs must be carefully taken into account while creating and developing electronic devices and networks [12]. Furthermore, the electrical conductivity of NMs is not well understood due to their smaller dimensions and novel mechanisms. Surface scattering, discrete bandgaps, Coulomb charging and tunneling and broadening are a few examples of the numerous processes. In addition, the enhanced perfection (i.e., fewer impurities, structural faults and dislocations) may have an impact on electrical conductivity of NMs. NMs may also store a substantial amount of energy compared to conventional materials because of their broad grain boundaries.

11.2.5 Optical behavior

Additionally, NMs are increasingly different from bulk forms due to their distinctive optical properties [13]. A couple of important properties are their effective charge and energy transmission spanning nanoscale distances and electrical carrier quantum confinement in NMs. The linear and nonlinear optical characteristics of NMs can be effectively changed by carefully managing the quantities of NMs as well as their surface chemistry.

11.3 Ceramic nanomaterials

Due to their superior physical and mechanical qualities compared to their bulk-sized equivalents, ceramic-based NMs are extensively sought-after for use in engineering. For example, magnesium aluminum has an improved melting point that enables its use in severe environments, and its improved optical qualities make it perfect for lenses [14]. Additionally, they are crucial for implants and tissue healing, and can be used as bioactive coatings on prosthesis [15]. Unfortunately, their toxicity is not well understood as that of many other NMs [16].

11.3.1 Synthesizing methods

Various synthetic techniques (Figure 11.1) are employed to make nanocoatings depending on the intended application. Other than customary techniques, like CVD and physical vapor deposition, more recent methods like sol–gel process and laser cladding are used to produce nanostructured coatings [18]. Several ways to make nanostructured coatings are listed in Figure 11.1. Substrates must also be prepared prior to the application of coatings, which requires actions like chemical modification cleaning. The quality of the coating applied to a substrate depends on the surface defects (pores and scratches), roughness, surface cleanliness (absence of contaminants, dirt and oxides/scales) and so on [19].

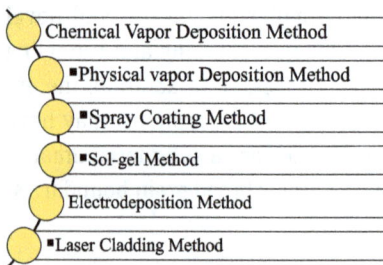

Chemical Vapor Deposition Method

∎Physical vapor Deposition Method

∎Spray Coating Method

∎Sol-gel Method

Electrodeposition Method

∎Laser Cladding Method

Figure 11.1: Nanocoating synthesis techniques: nanostructured coatings [17] (open access).

11.3.1.1 Chemical vapor deposition

The CVD technique entails the chemical dissociation of vaporous or gaseous materials close to a heated substrate surface, followed by coating application on a hot substrate surface. Metallic or ceramic compositions are coated using this technique. Typically, substrates are made of metallic or ceramic materials. It is important to consider the substrate's temperature because it can alter how well the coating sticks to the material [20]. Due to the intrinsic characteristics of the CVD method, multidirectional deposition of coating material on a substrate is conceivable [21]. This means that the approach may easily cover substrates with a range of sizes and complex geometries. Refractory materials are deposited on turbine blades via CVD, which is used in a wide range of industrial applications. The CVD technique is extortionate for coating big surfaces because it requires a slow rate of precipitation and higher substrate temperature. Depending on the chemical reaction, CVD procedures are divided into the following subcategories: atmospheric pressure CVD, plasma-assisted CVD, laser CVD, low-pressure CVD, plasma-enhanced CVD and atomic layer deposition (ALD) [22].

11.3.2 Types of ceramic nanomaterials

Due to their superior resistance to oxidation, corrosion and wear compared to metals, especially in high-temperature applications, ceramic nanocoatings are widely utilized in a variety of applications such as boiler components, engine valves, orthopedic implants and automotive body parts. They also possess thermal insulation and superior electrical properties [23]. In this section, the use of ceramic materials in nanocoatings is discussed.

11.3.2.1 Alumina (Al$_2$O$_3$) nanocoatings

Considering the high inherent resistance to mechanical wear and corrosion, as well as its low electrical and thermal conductivity, alumina (Al$_2$O$_3$) is a common material for ceramic coatings. Al$_2$O$_3$ nanocoatings' anticorrosion performance is influenced by the process used to create them. A layer of Al$_2$O$_3$ nanocoatings with a thickness varying from 10 to 50 nm was deposited on substrates made of 100Cr6 steel and Al2024-T3 aluminum. Due to superior nucleation of films compared to thermally enhanced and plasma-enhanced ALD nanocoatings, ALD nanocoatings were less porous. It was discovered that 10 nm thick nanocoating made by plasma-enhanced ALD clung to the substrate intact, whereas 10 nm thick nanocoating made by thermal-enhanced ALD had weak adhesion and came off the substrates. It was also discovered that the thickness of nanocoatings had an impact on their quality. The least porous Al$_2$O$_3$ nanocoating on both substrates was found to be 50 nm thick by both ALD procedures [24]. Because of good adhesion to substrates and their low porosity, thicker Al$_2$O$_3$ nanocoating (50 nm thickness) made using both ALD processes exhibited finer corrosion resistance [24]. Corrosion resistance is negatively impacted by the presence of porosity and a lack of strong substrate adherence for nanocoatings. Overall, it was determined that nanocoatings of 50 nm thick made by both ALD procedures offer the finest corrosion resistance and that nanocoatings formed by plasma-enhanced ALD provide superior corrosion resistance. Pretreatment is a primary process that can enhance the corrosion protection of nanocoatings. This was proven by hydrogen–argon plasma pretreatment of Al$_2$O$_3$ nanocoatings applied to steel using both plasma-enhanced and thermal-enhanced ALD, as well as its impact on corrosion performance of applied nanocoating [63]. The corrosion resistance of the nanocoatings created by both processes is improved by plasma pretreatment and increased pretreatment time. Due to the preeffect treatment on the substrate adhesion and porosity, the improvement in thermally enhanced ALD coatings was thought to be more significant. The elimination of heterogeneities through surface treatments of preannealing substrates in this manner before applying nanocoatings leads to better configuration of nanocoatings. The ability of nanocoatings to resist corrosion has been upgraded by preannealing copper substrates and then depositing 10–50 nm Al_2O_3 coatings by ALD [25]. It has been noted that the temperature of deposition affects the nanoceramic coatings' corrosive behavior during the ALD process. Al_2O_3 nanocoating was applied to 316L stainless steel by ALD at a temperature of 250 °C, and it was discovered that this coating had more corrosion resistance than coating applied at 160 °C. Higher temperature deposition increases the coating's sealing efficiency, which minimizes porosity and boosts corrosion resistance [26]. On the other hand, some carbon steels need lower deposition temperatures to prevent damaging impacts to their microstructure. To achieve efficient sealing and prevent corrosion, deposited Al_2O_3 nanocoating on 100Cr6 carbon steel at 160 °C using the ALD technique requires a >10 nm nanocoating thickness. The synthesis settings play a critical role in determining the qualities of the nanocoating, as may be inferred from the discussion above.

11.3.2.2 Titanium oxide (TiO_2) nanocoatings

Popular ceramic substance titanium oxide (TiO_2) is known for its resistance to mechanical abrasion [27] and corrosion, photocatalysis, self-cleaning property and protection against UV [28]. The size of the nanoceramic particles affects the corrosion resistance of ceramic nanocoatings. Studies on corrosion resistance of TiO_2 nanocoatings on carbon steel substrates revealed that corrosion resistance increases as the size of the nano-TiO_2 particles decreases [29]. Polarization and electrochemical impedance spectroscopy measurements of the corrosion rates of nanocoatings with 100, 150, 10 and 50 nm TiO_2 particle sizes in 1 M H_2SO_4 solution get deposited on the carbon steel surfaces, which demonstrated that corrosion is prevented by these nanocoatings. The physical connection between the coating and substrate surface was discovered to be dependent on the size of the nanoparticles. The coating–substrate interface bonding becomes better with smaller particle size. Because of lower O_2 and H_2O permeability into nanocoatings, resistance to corrosion of nanocoatings improves with the size reduction of nano-TiO_2 particles [29]. The anticorrosion efficacy of TiO_2 nanocoatings has been significantly enhanced with graphene oxide (GO) addition. A cast iron pipeline coated with a nanocomposite GO/TiO_2 ceramic coating demonstrated stunning 94% decrease in the rate of corrosion when compared with the surface of bare substrate in seawater. Mostly, this is caused by the coating's decreased porosity and capacitance. The thickness of TiO_2 nanocoatings is another element that affects how they behave when subjected to corrosion. A substrate with TiO_2 nanocoatings and un-coated AA2024 aluminum alloy substrate deposited at times of 80 and 40 s, respectively, are shown in images by scanning electron microscope (SEM) in Figure 11.2 [30]. With a current density that was one order of magnitude lower, the TiO_2 nanocoating with the thicker layer had the superior corrosion resistance which is achieved by a longer deposition time.

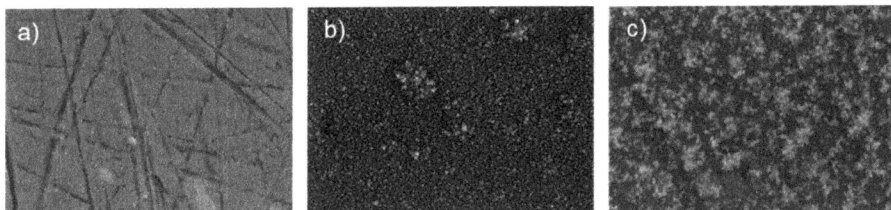

Figure 11.2: SEM images of TiO_2 nanocoating substrate [17] (open access).

11.3.2.3 Tantalum pentoxide (Ta_2O_5) nanocoatings

Ta_2O_5 in its ceramic state is the material of choice for creating nanocoatings which are corrosion-resistant [31]. It is highly robust [32] and resistant to chemical abrasion even in hostile environments. It is commonly used for mobile phones, automobile electronics and capacitors for high-speed machinery, due to its high dielectric constant (~25) [33].

Due to the development of a passive oxide stable layer, a β-Ta_2O_5 nanoceramic nanocoatings on a Ti–6Al–4V alloy substrate improved the resistance to corrosion in 3.5 wt% NaCl [34]. Chromium oxide (Cr_2O_3) and tantalum oxide (Ta_2O_5) nanocoatings on 100Cr6 steel substrate were compared for how well they prevented corrosion. Ta_2O_5-coated substrate nanocoating performed better than the substrate coated with Cr_2O_3 nanocoating [35]. Additionally, the mode of deposition had an impact on corrosion performance, with FCAD Ta_2O_5 nanocoatings showing roughly four times more corrosion-resistant than ALD-deposited nanocoating. Fictitious interfacial oxide layer produced by ALD coating promotes interface coating deterioration, interface voids and coating dissolution. The native oxide layer is removed by ion bombardment pre-etching substrate surface before actual oxide development (passive oxide coating) in the FCAD process [36].

11.3.2.4 Tantalum nitride (Ta₂N) nanocoatings

The corrosion properties of tantalum nitride (Ta$_2$N) that was coated over Ti–6Al–4V bipolar plates using the reactive sputter deposition [37] process were examined in a simulated polymer electrolyte membrane fuel cell environment with varying pH values and temperatures. In contrast to Ti–6Al–4V that was not coated, the Ta$_2$N-nanocoated substrate is substantially greater in terms of corrosion resistance at every given pH or temperature level. Corrosion resistance decreased as acidity increased [37]. As can be observed from Figure 11.3, Ta$_2$N nanocoatings displayed higher corrosion resistance in Ringer's physiological solution at 37 °C with lower I_{corr} values when compared to both pure Ta and bare Ti–6Al–4V, according to the research on nanocoatings on Ti–6Al–4V for biomaterial use [38].

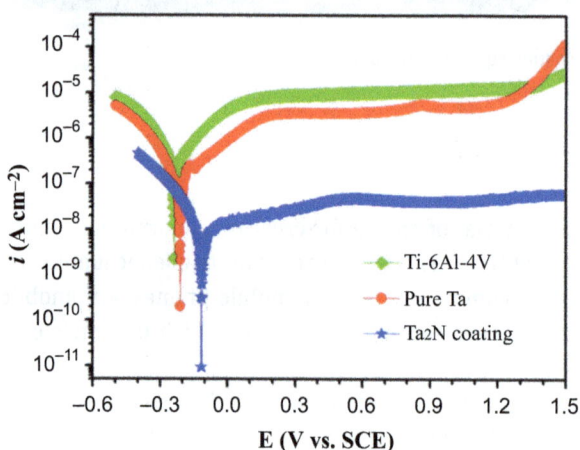

Figure 11.3: Polarization curves of Ta$_2$N coating, pure Ta and bare Ti–6Al–4V [17] (open access).

Ta$_2$O$_5$, Ta$_3$N$_5$ and TaON (tantalum oxynitride) nanocoatings made of tantalum were applied as coatings to 306 stainless steel. Hirpara et al. [39] demonstrated improved corrosion resistance by reducing corrosion current density as a result of the formation of passive film. When compared to samples of pure tantalum-coated stainless steel and bare stainless steel, the corrosion rates decreased by almost 50%. Among the Ta-based nanocoatings, TaON (tantalum oxynitride) showed the best corrosion resistance, followed by Ta$_2$O$_5$ (tantalum pentoxide) and Ta$_3$O$_5$ (tantalum nitride). TaON was said to have better corrosion resistance because of its hydrophobic properties, which were aided by its texture. Additionally, it was shown that the chemical, morphological and electrical characteristics of the deposited film had an impact on the anticorrosive qualities of the TaON nanocoating. This led to improved anticorrosive behavior since the TaON nanocoating greatly decreased the corrosion current density [39].

11.4 Anticorrosion application

Engineering systems can benefit from learning from natural systems [40]. This pattern is also evident in the evolution of coatings [41]. The development and design of coatings/surfaces for functions like enhancing interfacial adhesion, hydrophobicity, anticorrosion and reducing friction/wear are guided by a variety of natural system principles [42, 43]. Examples of biomimetic research on coatings for tribological purposes and corrosion are shown in this section.

By taking cues from nature, we can develop anticorrosion qualities through (i) surface texturing, (ii) hydrophobicity, (iii) microalloying and (iv) surface treatments. By creating superhydrophobic coatings with surface characteristics that resemble those of marine plants, corrosion resistance can be increased [44]. The shape of water-repellent plants can be used as an inspiration to create functional coatings that work well. The development of coatings that prevent the movement of corrosive media within coatings was motivated by the guided transit of fluids in many biological systems.

By combining the matrix of polyvinylidene fluoride with nano-SiO_2 and 2-mercaptobenzothiazole-loaded halloysite nanotubes, a hierarchical columnar structure with superhydrophobic and self-healing properties was created [45]. By limiting the flow of corrosive ions, two-dimensional materials like GOs and graphene have been used as fillers in coatings to delay or prevent corrosion [46]. A simple technique was used to create a zinc phosphate coating that mimicked the nepenthes pitcher plant and had superior corrosion resistance than the naked substrate [47]. It has been claimed that ultrathin graphene nanosheets placed between epoxy layers can successfully prevent galvanic corrosion by imitating the microstructures/nanostructures of nacre and mussels [48]. Anticorrosion coatings doped with ion-selective structures and resins inspired by marine mangrove leaves [49] have demonstrated the capacity to control corrosive medium migration at the coating metal interface.

Each of the aforementioned methods for depositing thin films on a substrate's surface has an impact on the uniformity and surface characteristics including fracture toughness, strength and ductility [50]. Every technique has advantages and disadvantages; thus, it is important to consider all of the processing factors before selecting the best one. In order to get the best surface coverage of the nanocoating in terms of smoothness, uniformity, crack-free surfaces and adhesion, the process should be applied under optimal conditions. For instance, dip coating can coat complex structures and is affordable, but may experience necessary high sintering temperatures and thermal expansion mismatch. Pulsed laser deposition and hot pressing can provide uniform, dense coatings, but they both have dip drawbacks. Processes such as sputter coating and thermal spraying may result in an amorphous structure as a result of quick cooling. The recommended method is sol–gel because it is a reasonably inexpensive coating and has low pressing temperatures, but it needs a regulated processing environment [51] and expensive raw materials. The conditions and processing steps of various nanocoating processes should therefore be optimized, as multistep technologies are not appealing in the industrial market.

In some situations, nanocoatings might not be effective as protective surfaces. Due to the high density of their grain boundaries, which allow for quick diffusion routes for passivated ions and improved adherence of the protective oxide layer to the substrate surface, nanocoatings are efficient physical barriers in high-temperature applications [52]. However, because there are more anodic sites due to the greater grain boundary fraction, the surface is more vulnerable to corrosion attack. Additionally, by incorporating into the vacancies, dislocations and grain/interphase barriers, nanocoatings create a protective structure. Due to the rapid diffusion of passivating ions, these characteristics have the advantage of generating a more effective passivation layer that is more effective. On the other hand, the agglomeration of these nanoscale materials may result from rapid diffusion of hostile ions, which leads to uneven surfaces and raises the likelihood of formation of active sites, thus reducing resistance from corrosion [53]. Such a disagreement highlights the need to investigate the corrosive behavior of each nanocoating while considering all relevant environmental factors.

One element of nanocoating is on the nanoscale. The relatively small sizes of the particle utilized in this nanocoating make it more effective in filling voids and preventing corrosive element from diffusing into substrate's surface. Additionally, high density of the grain boundaries in nanocoatings offers enhanced adhesive qualities, extending the coating's life span [54]. Nanocoatings offer improved mechanical and electrical characteristics that make them harder, stronger and more resistant to conditions that cause corrosion and wear [55, 56]. Self-healing [57], self-cleaning [56], wear resistance [58] and excellent scratch are only a few of the new features that nanocoating technology has added to paints, which has had a significant impact on their growth. It also made alternatives to the available chromium hazardous coating [59, 60]. Similar to this, smart nanocoatings have significant advantages in decreasing the impacts of biofouling and corrosion. They are designed to release controlled amount of inhibitor in response to environmental stimuli including humidity, pH, heat, coating deformation, stress and electromagnetic radiation in order to mend and heal flaws and damages [61, 62]. The corrosion behavior of conventional coatings with thickness or microsized particle would be different from that of nanocoating [63]. For instance, applying a nanothick zinc coating can solve issues with poor weldability and obtain a specular surface after painting [63]. Because of the remarkable qualities that nanocoatings possess, they are employed in products that are used every day, including apparel, laptops, cell phones and eyeglasses. They are utilized in tiles, flooring, windows, walls, air filters, paints and other aspects of construction. These appliances are made wear-resistant, scratch-resistant, flame-resistant, anti-graffiti, self-cleaning, corrosion-resistant and electrically conductive as nanolayer is used in them. They are suitable as photovoltaic materials and have good adhesion, antifogging, optical clarity and antifouling qualities [64]. Metallic nanocoating is employed in the biomedical industries to change the surface quality as necessary. In addition to additional secondary uses including biocompatibility and medication administration, they are principally used in the medical sector for etch surface covering, protection and anticorrosion activities [65]. Nanocoatings are used in many other industries, including the automotive sector, military, environment and energy efficiency, because of the properties listed above. Corrosion prevention is greatly aided by the inclusion of nanoparticles on the matrix of the coating. Silica nanoparticles after treatment are used to increase dispersibility in the polymeric coating. It was discovered that corrosion resistance increased and up to 5% SiO_2 concentration before loading corrosion began. Later that amount of concentration, the nanoparticles begin to aggregate, and the binding toughness between the substrate and coating weakens [66], according to SEM pictures. Coated nanocomposites have multiple ways to minimize corrosion. In addition to the precise occurrence of electrical conductivity within the polymeric matrix, the resistance effect caused by the creation of a yielding film on the facet of the coated nanocomposite is another result that offers corrosion prevention. Moreover, low overpotential areas and a reduction in surface corrosion reactions can be provided by oxygen reduction on the polymer surface. A potential corrosion apparatus on the top of a nanocoating is displayed schematically in Figure 11.4.

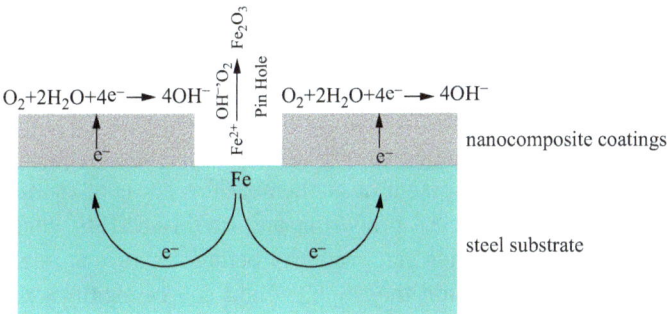

Figure 11.4: Anticorrosion process by nanocomposites on steel substrate.

11.5 Challenges

Practical obstacles exist in the production and evolution of nanostructured coatings for required tribological performance and anticorrosion: (i) the proper nanocoating material selection, tailored to a certain substrate and operating condition (e.g., adhesion to the substrate, which is more difficult for materials with complicated geometries, change in size and ease of deposition); (ii) choosing the best reinforcing for a particular coating type and purpose (e.g., thermal discrepancy matrix material and mechanical compatibility); (iii) chemical complexity involved in the synthesis of nanocoatings; (iv) the efficiency of creating and implementing nanocoatings in terms of cost and time; (v) capacity to coat vast surfaces using microcoats; and (vi) performance and upkeep throughout the long haul.

11.6 Conclusions

This chapter renders a deep examination of synthesis, corrosion behavior and tribological performance of nanocomposite coatings and nanostructured coatings (metallic coatings, ceramic coatings and metal and polymer matrices with nanocomposite coatings). Given that many factors, such as composition, synthesis method, coating material, grain size, operating environment, additives, contact conditions and processing parameters, affect the properties and coating performance, it will be a difficult undertaking for developing coatings for tribological applications and corrosion. The creation of efficient coatings is made even more difficult by problems including the choice of chemical complexity, coating material/reinforcements and cost and time efficiency in coating synthesis. Utilization of biomimetic coverings in tribological and anticorrosion applications has shown great promise. The performance of newly created nanostructured coatings and nanocomposite coatings has been great on a lab scale, but their use in practical settings has not yet been accomplished.

Numerous technical fields, including maritime, space/aerospace, robotics, automotive medicine (like orthopedic and dentistry), architecture, defense, sports and energy systems, benefit from usage of nanostructured layers. The critical concerns that are needed to be resolved to realize the full potential of coatings for application are as follows: (i) interfacial incremental adhesion between coatings and substrates; (ii) introduction of low-cost coating preparation processes; (iii) worthwhile design lessons from natural systems; and (iv) the development of mathematical replicas of coating processes and their influence on tribological properties and corrosion. This would make it easier to optimize coating qualities and processes before they are applied. The evolution of nanostructured layers and their applications can be significantly aided by additive manufacturing methods [21].

References

[1]   Palit, S., Hussain, C.M., Green nanomaterials: A sustainable perspective, Advanced Structured Materials, 2020, 126, doi: https://doi.org/10.1007/978-981-15-3560-4_2.
[2]   Keçili, R., Arli, G., Hussain, C.M., Future of analytical chemistry with graphene, in Comprehensive Analytical Chemistry, Vol. 91, 2020, doi: https://doi.org/10.1016/bs.coac.2020.09.003.
[3]   Awad, A.M., Jalab, R., Benamor, A., Nasser, M.S., Ba-Abbad, M.M., El-Naas, M., Mohammad, A.W., Adsorption of organic pollutants by nanomaterial-based adsorbents: An overview, Journal of Molecular Liquids, 2020, 301, 112335.
[4]   Barhoum, A., García-Betancourt, M.L., Physicochemical characterization of nanomaterials: Size, morphology, optical, magnetic, and electrical properties, in Emerging Applications of Nanoparticles and Architecture Nanostructures, Elsevier, 2018, 279–304.
[5]   Yurkov, G.Y., Fionov, A.S., Koksharov, Y.A., Koleso, V.V., Gubin, S.P., Electrical and magnetic properties of nanomaterials containing iron or cobalt nanoparticles, Inorganic Materials, 2007, 43, 834–844.

[6]   Sharma, V., Optical properties of tin oxide nanomaterials, in Tin Oxide Materials, Elsevier, 2020, 61–99.

[7]   Wang, X.H., Huang, L.Q., Niu, L.J., Li, R.B., Fan, D.H., Zhang, F.B., Chen, Z.W., Wang, X., Guo, Q.X., The impacts of growth temperature on morphologies, compositions and optical properties of Mg-doped ZnO nanomaterials by chemical vapor deposition, Journal of Alloys and Compounds, 2015, 622, 440–445.

[8]   Das, P., Fatehbasharzad, P., Colombo, M., Fiandra, L., Prosperi, D., Multifunctional magnetic gold nanomaterials for cancer, Trends in Biotechnology, 2019, 37(9), 995–1010.

[9]   Felix, L.C., Ede, J.D., Snell, D.A., Oliveira, T.M., Martinez-Rubi, Y., Simard, B., Luong, J.H.T., Goss, G.G., Physicochemical properties of functionalized carbon-based nanomaterials and their toxicity to fishes, Carbon, 2016, 104, 78–89.

[10]   Sadik, O.A., Du, N., Kariuki, V., Okello, V., Bushlyar, V., Current and emerging technologies for the characterization of nanomaterials, ACS Sustainable Chemistry & Engineering, 2014, 2(7), 1707–1716.

[11]   Mwafy, E.A., Hasanin, M.S., Mostafa, A.M., Cadmium oxide/TEMPO-oxidized cellulose nanocomposites produced by pulsed laser ablation in liquid environment: Synthesis, characterization, and antimicrobial activity, Optics & Laser Technology, 2019, 120, 105744.

[12]   Abu-Thabit, N.Y., Nanomaterials for flexible transparent conductive films and optoelectronic devices, in Advances in Smart Coatings and Thin Films for Future Industrial and Biomedical Engineering Applications, Elsevier, 2020, 619–643.

[13]   Shih, P.-H., Do, T.-N., Gumbs, G., Lin, M.-F., Electronic and optical properties of doped graphene, Physica E: Low-Dimensional Systems and Nanostructures, 2020, 118, 113894.

[14]   Li, C., Sun, W., Lu, Z., Ao, X., Li, S., Ceramic nanocomposite membranes and membrane fouling: A review, Water Research, 2020, 175, doi: https://doi.org/10.1016/j.watres.2020.115674.

[15]   Peter, M., Binulal, N.S., Nair, S.V., Selvamurugan, N., Tamura, H., Jayakumar, R., Novel biodegradable chitosan–gelatin/nano-bioactive glass ceramic composite scaffolds for alveolar bone tissue engineering, Chemical Engineering Journal, 2010a, 158(2), 353–361.

[16]   Park, E.-J., Park, K., Oxidative stress and pro-inflammatory responses induced by silica nanoparticles in vivo and in vitro, Toxicology Letters, 2009, 184(1), 18–25.

[17]   Farooq, S.A., Raina, A., Mohan, S., Arvind Singh, R., Jayalakshmi, S., Irfan Ul Haq, M., Nanostructured coatings: Review on processing techniques, corrosion behaviour and tribological performance, Nanomaterials, 2022, 12(8), 1323.

[18]   Gu, Y., Xia, K., Wu, D., Mou, J., Zheng, S., Technical characteristics and wear-resistant mechanism of nano coatings: A review, Coatings, 2020, 10(3), doi: https://doi.org/10.3390/coatings10030233.

[19]   Aliofkhazraei, M., Aliofkhazraei, M., Synthesis, processing and application of nanostructured coatings, Nanocoatings: Size Effect in Nanostructured Films, 2011, 1–28.

[20]   Makhlouf, A.S.H., Current and advanced coating technologies for industrial applications, in Nanocoatings and Ultra-thin Films, Elsevier, 2011, 3–23.

[21]   Behera, A., Mallick, P., Mohapatra, S.S., Nanocoatings for anticorrosion: An introduction, in Corrosion Protection at the Nanoscale, Elsevier, 2020, 227–243.

[22]   Shishkovsky, I.V., Lebedev, P.N., Chemical and physical vapor deposition methods for nanocoatings, in Nanocoatings and Ultra-thin Films, Elsevier, 2011, 57–77.

[23]   Callister Jr, W.D., Rethwisch, D.G., Fundamentals of Materials Science and Engineering: An Integrated Approach, John Wiley & Sons, 2020.

[24]   Koyama, S., Haniu, H., Osaka, K., Koyama, H., Kuroiwa, N., Endo, M., Kim, Y.A., Hayashi, T., Medical application of carbon-nanotube-filled nanocomposites: The microcatheter, Small, 2006, 2(12), doi: https://doi.org/10.1002/smll.200500416.

[25]   Krätschmer, W., Lamb, L.D., Fostiropoulos, K., Huffman, D.R., Solid C60: A new form of carbon, Nature, 1990, 347(6291), doi: https://doi.org/10.1038/347354a0.

[26] Magrez, A., Kasas, S., Salicio, V., Pasquier, N., Seo, J.W., Celio, M., Catsicas, S., Schwaller, B., Forró, L., Cellular toxicity of carbon-based nanomaterials, Nano Letters, 2006, 6(6), doi: https://doi.org/10.1021/nl060162e.

[27] Poland, C.A., Duffin, R., Kinloch, I., Maynard, A., Wallace, W.A.H., Seaton, A., Stone, V., Brown, S., MacNee, W., Donaldson, K., Carbon nanotubes introduced into the abdominal cavity of mice show asbestos-like pathogenicity in a pilot study, Nature Nanotechnology, 2008, 3(7), doi: https://doi.org/10.1038/nnano.2008.111.

[28] Silvestre, C., Duraccio, D., Cimmino, S., Food packaging based on polymer nanomaterials, Progress in Polymer Science (Oxford), 2011, 36(12), doi: https://doi.org/10.1016/j.progpolymsci.2011.02.003.

[29] Lucarelli, M., Gatti, A.M., Savarino, G., Quattroni, P., Martinelli, L., Monari, E., Boraschi, D., Innate defence functions of macrophages can be biased by nano-sized ceramic and metallic particles, European Cytokine Network, 2004, 15(4).

[30] Peter, M., Binulal, N.S., Nair, S.V., Selvamurugan, N., Tamura, H., Jayakumar, R., Novel biodegradable chitosan-gelatin/nano-bioactive glass ceramic composite scaffolds for alveolar bone tissue engineering, Chemical Engineering Journal, 2010b, 158(2), doi: https://doi.org/10.1016/j.cej.2010.02.003.

[31] Peraldo, L., Fairbrother, F. The Chemistry of Niobium and Tantalum, 1969.

[32] Rahmati, B., Sarhan, A.A.D., Zalnezhad, E., Kamiab, Z., Dabbagh, A., Choudhury, D., Abas, W., Development of tantalum oxide (Ta-O) thin film coating on biomedical Ti-6Al-4V alloy to enhance mechanical properties and biocompatibility, Ceramics International, 2016, 42(1), 466–480.

[33] Habibi, M.H., Mokhtari, R., Novel sulfur-doped niobium pentoxide nanoparticles: Fabrication, characterization, visible light sensitization and redox charge transfer study, Journal of Sol-Gel Science and Technology, 2011, 59(2), 352–357, doi: https://doi.org/10.1007/s10971-011-2510-z.

[34] Hu, W., Xu, J., Lu, X., Hu, D., Tao, H., Munroe, P., Xie, Z.-H., Corrosion and wear behaviours of a reactive-sputter-deposited Ta2O5 nanoceramic coating, Applied Surface Science, 2016, 368, 177–190, doi: https://doi.org/https://doi.org/10.1016/j.apsusc.2016.02.014.

[35] Díaz, B., Świątowska, J., Maurice, V., Pisarek, M., Seyeux, A., Zanna, S., Tervakangas, S., Kolehmainen, J., Marcus, P., Chromium and tantalum oxide nanocoatings prepared by filtered cathodic arc deposition for corrosion protection of carbon steel, Surface and Coatings Technology, 2012, 206(19), 3903–3910, doi: https://doi.org/https://doi.org/10.1016/j.surfcoat.2012.03.048.

[36] Díaz, B., Świątowska, J., Maurice, V., Seyeux, A., Härkönen, E., Ritala, M., Tervakangas, S., Kolehmainen, J., Marcus, P., Tantalum oxide nanocoatings prepared by atomic layer and filtered cathodic arc deposition for corrosion protection of steel: Comparative surface and electrochemical analysis, Electrochimica Acta, 2013, 90, 232–245, doi: https://doi.org/https://doi.org/10.1016/j.electacta.2012.12.007.

[37] Jin Ma, J., Xu, J., Jiang, S., Munroe, P., Xie, Z.-H., Effects of pH value and temperature on the corrosion behavior of a Ta2N nanoceramic coating in simulated polymer electrolyte membrane fuel cell environment, Ceramics International, 2016a, 42(15), 16833–16851, doi: https://doi.org/https://doi.org/10.1016/j.ceramint.2016.07.175.

[38] Cheng, J., Xu, J., Liu, L.L., Jiang, S., Electrochemical corrosion behavior of Ta2n nanoceramic coating in simulated body fluid, Materials, 2016, 9(9), 772.

[39] Hirpara, J., Chawla, V., Chandra, R., Anticorrosive behavior enhancement of stainless steel 304 through tantalum-based coatings: Role of coating morphology, Journal of Materials Engineering and Performance, 2021, 30(3), 1895–1905, doi: https://doi.org/10.1007/s11665-021-05542-5.

[40] Bhushan, B., Biomimetics: Lessons from nature – an overview, Philosophical Transactions of the Royal Society A: Mathematical, Physical and Engineering Sciences, 2009, 367(1893), 1445–1486, doi: https://doi.org/10.1098/rsta.2009.0011.

[41] Jabbari, E., Kim, D.-H., Lee, L.P., Handbook of Biomimetics and Bioinspiration: Biologically-driven Engineering of Materials, Processes, Devices, and Systems, World Scientific, 2014.

[42] Evans, H.B., Hamed, A.M., Gorumlu, S., Doosttalab, A., Aksak, B., Chamorro, L.P., Castillo, L., Engineered bio-inspired coating for passive flow control, Proceedings of the National Academy of Sciences of the United States of America, 2018, 115(6), 1210–1214, doi: https://doi.org/10.1073/pnas.1715567115.

[43] Smith, G.M., Resnick, M., Flynn, K., Dwivedi, G., Sampath, S., Nature inspired, multi-functional, damage tolerant thermal spray coatings, Surface and Coatings Technology, 2016, 297, 43–50, doi: https://doi.org/https://doi.org/10.1016/j.surfcoat.2016.04.047.

[44] Xu, J., Cai, Q., Lian, Z., Yu, Z., Ren, W., Yu, H., Research progress on corrosion resistance of magnesium alloys with bio-inspired water-repellent properties: A review, Journal of Bionic Engineering, 2021, 18(4), 735–763, doi: https://doi.org/10.1007/s42235-021-0064-5.

[45] Yang, X., Tian, L., Wang, W., Fan, Y., Sun, J., Zhao, J., Ren, L., Bio-inspired superhydrophobic self-healing surfaces with synergistic anticorrosion performance, Journal of Bionic Engineering, 2020, 17(6), 1196–1208, doi: https://doi.org/10.1007/s42235-020-0094-4.

[46] Chang, K.C., Ji, W.F., Lai, M.C., Hsiao, Y.R., Hsu, C.H., Chuang, T.L., Wei, Y., Yeh, J.M., Liu, W.R., Synergistic effects of hydrophobicity and gas barrier properties on the anticorrosion property of PMMA nanocomposite coatings embedded with graphene nanosheets, Polymer Chemistry, 2014, 5 (3), 1049–1056, doi: https://doi.org/10.1039/c3py01178j.

[47] Xiang, T., Zheng, S., Zhang, M., Sadig, H.R., Li, C., Bioinspired slippery zinc phosphate coating for sustainable corrosion protection, ACS Sustainable Chemistry & Engineering, 2018, 6(8), 10960–10968, doi: https://doi.org/10.1021/acssuschemeng.8b02345.

[48] Zhu, X., Zhao, H., Wang, L., Xue, Q., Bioinspired ultrathin graphene nanosheets sandwiched between epoxy layers for high performance of anticorrosion coatings, Chemical Engineering Journal, 2021, 410, 128301, doi: https://doi.org/https://doi.org/10.1016/j.cej.2020.128301.

[49] Cui, M., Wang, P.Y., Wang, Z., Wang, B., Mangrove inspired anti-corrosion coatings, Coatings, 2019, 9 (11), doi: https://doi.org/10.3390/coatings9110725.

[50] Agarwala, V., Chandra Agarwala, R., Sunder Daniel, B.S., Development of nanograined metallic materials by bulk and coating techniques, Synthesis and Reactivity in Inorganic, Metal-Organic, and Nano-Metal Chemistry, 2006, 36(1), 3–16, doi: https://doi.org/10.1080/15533170500471128.

[51] Salam Hamdy, A., Corrosion protection performance via nano-coatings technologies, Recent Patents on Materials Science, 2010, 3(3), 258–267.

[52] Gao, W., Li, Z., Nano-structured alloy and composite coatings for high temperature applications, Materials Research, 2004, 7, 175–182.

[53] Saji, V.S., Cook, R.M., Corrosion Protection and Control using Nanomaterials, Elsevier, 2012.

[54] Jones, D.A., Principles and prevention, Corrosion, 1996, 2, 168.

[55] Schuh, C.A., Nieh, T.G., Iwasaki, H., The effect of solid solution W additions on the mechanical properties of nanocrystalline Ni, Acta Materialia, 2003, 51(2), 431–443.

[56] Sriraman, K.R., Strauss, H.W., Brahimi, S., Chromik, R.R., Szpunar, J.A., Osborne, J.H., Yue, S., Tribological behavior of electrodeposited Zn, Zn–Ni, Cd and Cd–Ti coatings on low carbon steel substrates, Tribology International, 2012, 56, 107–120.

[57] Andreatta, F., Aldighieri, P., Paussa, L., Di Maggio, R., Rossi, S., Fedrizzi, L., Electrochemical behaviour of ZrO2 sol–gel pre-treatments on AA6060 aluminium alloy, Electrochimica Acta, 2007, 52(27), 7545–7555, doi: https://doi.org/https://doi.org/10.1016/j.electacta.2006.12.065.

[58] Wang, Y., Zhang, L., Hu, Y., Li, C., Comparative study on optical properties and scratch resistance of nanocomposite coatings incorporated with flame spray pyrolyzed silica modified via in-situ route and ex-situ route, Journal of Materials Science & Technology, 2016, 32(3), 251–258, doi: https://doi.org/https://doi.org/10.1016/j.jmst.2015.11.008.

[59] Hibbard, G., Aust, K.T., Palumbo, G., Erb, U., Thermal stability of electrodeposited nanocrystalline cobalt, Scripta Materialia, 2001, 44(3), 513–518, doi: https://doi.org/https://doi.org/10.1016/S1359-6462(00)00628-X.

[60] Jin Ma, J., Xu, J., Jiang, S., Munroe, P., Xie, Z.-H., Effects of pH value and temperature on the corrosion behavior of a Ta2N nanoceramic coating in simulated polymer electrolyte membrane fuel cell environment, Ceramics International, 2016b, 42(15), 16833–16851, doi: https://doi.org/https://doi.org/10.1016/j.ceramint.2016.07.175.

[61] McGee, J.D., Smith, I.I.T.S., Bammel, B.D., Bryden, T.R., Release on Demand Corrosion Inhibitor Composition, Google Patents, 2012.

[62] Rahmani, K., Jadidian, R., Haghtalab, S., Evaluation of inhibitors and biocides on the corrosion, scaling and biofouling control of carbon steel and copper–nickel alloys in a power plant cooling water system, Desalination, 2016, 393, 174–185, doi: https://doi.org/https://doi.org/10.1016/j.desal.2015.07.026.

[63] Youssef, K.M., Koch, C.C., Fedkiw, P.S., Improved corrosion behavior of nanocrystalline zinc produced by pulse-current electrodeposition, Corrosion Science, 2004, 46(1), 51–64, doi: https://doi.org/https://doi.org/10.1016/S0010-938X(03)00142-2.

[64] Beyene, F.G., A review on nanocoating of metallic structures to improve hardness and maintaining toughness, I-Manager's Journal on Material Science, 2016, 4(1), 32.

[65] Mahapatro, A., Bio-functional nano-coatings on metallic biomaterials, Materials Science and Engineering: C, 2015, 55, 227–251, doi: https://doi.org/https://doi.org/10.1016/j.msec.2015.05.018.

[66] Chen, L., Song, R.G., Li, X.W., Guo, Y.Q., Wang, C., Jiang, Y., The improvement of corrosion resistance of fluoropolymer coatings by SiO2/poly(styrene-co-butyl acrylate) nanocomposite particles, Applied Surface Science, 2015, 353, 254–262, doi: https://doi.org/https://doi.org/10.1016/j.apsusc.2015.06.148.

Khasan Berdimuradov, Elyor Berdimurodov*, Ilyos Eliboev,
Nurbek Umirov, Bakhtiyor Borikhonov, Abduvali Kholikov,
Khamdam Akbarov

Chapter 12
Smart–hybrid nanomaterials in corrosion prevention

Abstract: Currently, metallic materials are protected with smart–hybrid nanomaterials. There are three types of smart–hybrid nanomaterials: self-cleaning, self-healing and superhydrophobic. The smart–hybrid nanomaterials were made from inorganic-, organic-, polymer-, silica-, epoxy- and carbon-based materials. Some self-healing smart–hybrid nanomaterials were prepared from the inorganic nanocoatings. These types of nanomaterials are low cost and more effective in corrosion protection. The metal surface was covered with these types of nanocoatings in chemical and electrochemical ways. The following metals were used in this self-healing of smart–hybrid nanomaterials: zinc-, nickel-, cerium-, potassium-, phosphate-, vanadium-, aluminum-, chromium-, zirconium- and molybdenum-rich layers. The metal materials were protected by using nanocoatings. Smart and hybrid nanomaterials were a new trend in corrosion protection. Smart–hybrid nanomaterials have some advantages in metal protection: they form the barrier oxide layer on the metal surface. The organic compounds and silica-based nanomaterials were most used in corrosion protection. These nanocoatings contained more halogen, alkoxy, hydroxyl, carboxyl and hydrolyzable functional groups. These types of materials are also named organosilane nanomaterials. The protective coatings are missed with organosilane to form the hybrid nanomaterials.

Keywords: Corrosion inhibitor, smart nanomaterials, self-healing, self-cleaning, hybrid nanomaterials, nanocoatings, metal protection

*Corresponding author: Elyor Berdimurodov, Faculty of Chemistry, National University of Uzbekistan, Tashkent 100034, Uzbekistan, e-mail: elyor170690@gmail.com
Khasan Berdimuradov, Faculty of Industrial Viticulture and Food Production Technology, Shahrisabz branch of Tashkent Institute of Chemical Technology, Shahrisabz 181306, Uzbekistan
Ilyos Eliboev, Abduvali Kholikov, Khamdam Akbarov, Faculty of Chemistry, National University of Uzbekistan, Tashkent 100034, Uzbekistan
Nurbek Umirov, Bakhtiyor Borikhonov, Faculty of Chemistry–Biology, Karshi State University, Karshi 130100, Uzbekistan

https://doi.org/10.1515/9783111071756-012

12.1 Introduction

Corrosion protection is an important task in the chemical and gas oil industries. The corrosion protection required smart nanomaterials. Currently, metallic materials are protected with smart–hybrid nanomaterials [1–3]. There are three types of smart–hybrid nanomaterials: self-cleaning, self-healing and superhydrophobic. The smart–hybrid nanomaterials were made from inorganic-, organic-, polymer-, silica-, epoxy- and carbon-based materials. These types of smart–hybrid nanomaterials have various good properties. For example, polymer-, silica- and epoxy-based nanomaterials have excellent mechanical and thermal stability. The smart–hybrid nanomaterials were also named self-assembled nanophase particles because the monomers were simultaneously self-assembled in the coating processes [3–5]. The self-cleaning type of smart–hybrid nanomaterials contained more supramolecular systems. The biological membranes, liposomes, micelles, biomolecular condensates, liquid crystals and colloids were modern supramolecular systems. Cyclodextrin is also widely used in the self-cleaning type of smart–hybrid nanomaterials [6]. The cyclodextrin easily formed a host–guest system with the organic corrosion inhibitors. Then the formed complex was added to the protective coatings, and these coatings are used in metal protection. In self-cleaning processes, the corrosion products are cleaned with corrosion inhibitors. In comparison to self-healing processes, the corroded parts of metal materials were self-protected with the organic compound, which was attached to the inner cavity of the cyclodextrin supramolecular system [7–9].

The basic properties of smart–hybrid nanomaterials are identified in Figure 12.1. Good smart–hybrid nanomaterials have good electrical and magnetic properties, such as good electromagnetic, antistatic, shield, dielectric, insulating and conductive properties [10, 11].

On the other hand, smart–hybrid nanomaterials also have good optical performances: photochromic, antireflection, photocatalytic and photoluminescence. These properties promote the corrosion protection of metallic materials under ultralight corrosion [12].

The mechanical properties also play a good role in smart–hybrid nanomaterials. For example, these types of materials show good lubrication, hardness, abrasion and wear. The thermal and chemical stability of smart–hybrid nanomaterials is also high [13, 14].

The following chemical and physical performances made the smart–hybrid nanomaterials to become high corrosion-protective materials: fire-retardant, antimicrobial, antifouling, self-cleaning, hydrophobic, hydrophilic, thermal barrier and corrosion protection [15].

The inorganic and organic materials were modified to form smart–hybrid nanomaterials. These nanomaterials can efficiently protect the metal surface in aggressive acidic, alkaline and other aggressive solutions. The greenness of these nanomaterials is also high because natural polymers or compounds were performed to prepare the smart–hybrid nanomaterials [12, 16].

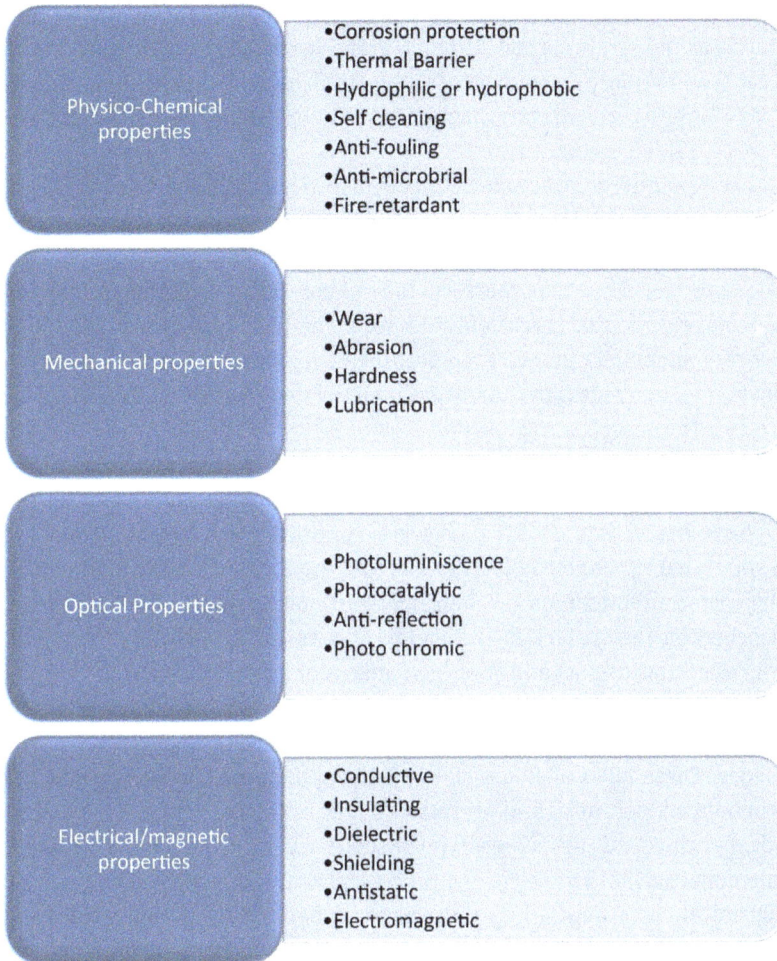

Figure 12.1: Basic properties of smart–hybrid nanomaterials [15].

12.2 Main part

12.2.1 Sol–gel type of smart–hybrid nanomaterials in corrosion protection

The sol–gel type of smart–hybrid nanomaterials was prepared from the polycondensation reaction of precursors in the liquid phase. The colloidal particles were prepared by the sol–gel methods. The sol–gel type of smart–hybrid nanomaterials is similar to the colloidal sols. The hybrid–smart type of nanomaterials for corrosion protection was prepared by the sol–gel methodology. This methodology has several advantages, such as the synthesis source being green materials, the synthesis procedure being environmentally friendly, does not requiring expensive chemicals or equipment, the synthesis temperature being near-room temperature, waste free and easy operation [17, 18] (Figure 12.2).

On the other hand, the sol–gel type of smart–hybrid nanomaterials has high specific surface area and specific porosity. These properties promote high corrosion protection efficiency. The corrosion inhibitors are absorbed on the surface of high-porosity nanomaterials. The smart performance of sol–gel-type nanomaterials depends on the concentration of corrosion inhibitors. The high surface area and high surface porosity of nanomaterials are good adsorbents for corrosion inhibitors. When the metal surface was corroded, the corrosion inhibitors were released to protect the metal surface by adsorption [19, 20].

Some sol–gel-type nanomaterials were synthesized at low temperatures. These processes require the degradation of the entrapped species and minimal thermal volatilization. Cr(III), Ce(IV) and Ce(III) ions were modified with the surface of the sol–gel-type smart–hybrid nanomaterials to increase the corrosion protection performance. In the corrosion–inhibition mechanism, Cr(III), Ce(IV) and Ce(III) ions are effectively adsorbed on the surface of the metal. As a result, the metal surface was insulated from the water contact and corrosion attacks of ions [21].

The sol–gel-type nanomaterials were also composited with the sacrificial metal pigments (magnesium, aluminum, zinc and their alloy particles) to enhance the protection efficiency. These types of smart materials form the protective layer, which is more stable in high temperatures and aggressive acidic solutions.

In chemical engineering, the sol–gel type of smart–hybrid nanomaterials was deposited on the metal surface by the sol–gel process, plasma spraying, electrochemical deposition, chemical vapor deposition, physical vapor deposition and other mechanical methods. The metallic materials were defended by the formation of the excellent layer of a sol–gel type of smart–hybrid nanomaterials [22].

12.2.2 Self-healing of smart–hybrid nanomaterials in the corrosion protection

The metal materials were protected by using nanocoatings. Smart and hybrid nanomaterials were a new trend in corrosion protection. This is due to the smart–hybrid nanomaterials having some advantages in metal protection: they form the barrier oxide layer on the metal surface. The size thickness of this film is around a few nanometers. These properties are mainly responsible for the anticorrosion performance of nanocoatings. The self-healing performance of nanocoatings is as follows [24] (Figure 12.3):

(a)

(b)

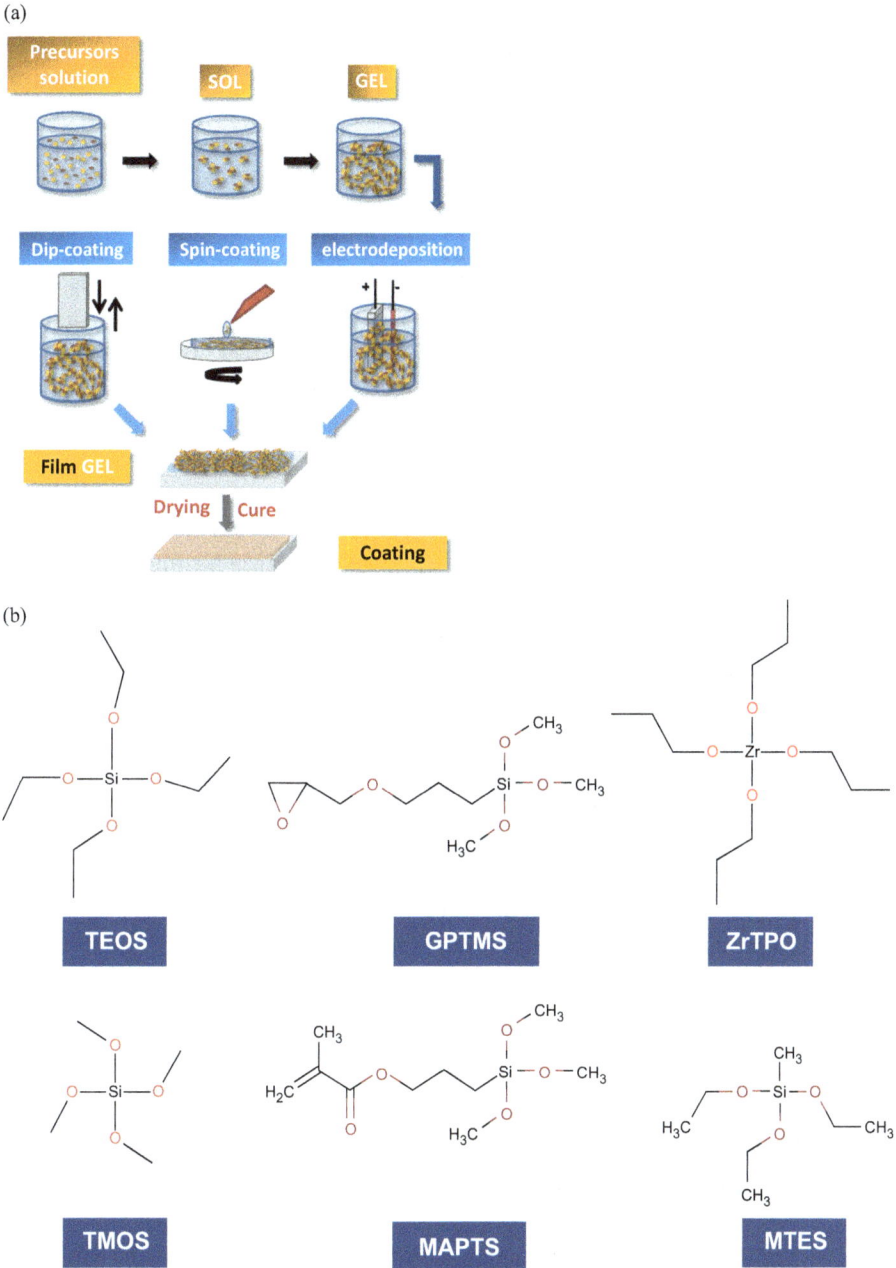

Figure 12.2: (a) Preparation methods [23] and (b) some examples of the sol–gel type of smart–hybrid nanomaterials [22].

(i)   The smart–hybrid nanomaterials contained the CD-based containers. The main aim of containers is that they capture large amounts of corrosion inhibitors. The supramolecular system or host–guest systems were formed between the corrosion inhibitor and Cyclodextrin (CD) inner cavity. When the surface of the metal was cracked or distracted, the corrosion inhibitor was released, thereby protecting the metal surface by good chemical or electrochemical interactions [25, 26].

(ii) The underlying substrate was passivated by the noble metal based on the smart–hybrid nanomaterials. The noble metals and polymer coatings were mixed to form the noble metal-based smart–hybrid nanomaterials. These materials were used to protect the metallic materials. The noble-based metals or their compositions are good passivates for the metal surface. As a result, the oxidation processes were depleted dramatically [27, 28].

(iii) The corrosion processes have occurred in the cathodic and anodic regions. The smart–hybrid nanomaterials contain more corrosion inhibitors, which are mainly responsible for corrosion protection.

(iv) The electrochemical reactions, such as anodic or cathodic reactions, are high rate in the corrosion processes because the electrical resistance for the electrochemical reactions is very low. However, the electrical resistance considerably rose with the formation of the protective layer of smart–hybrid nanomaterials on the metal surface.

(v)   The smart–hybrid nanomaterials promote cathodic corrosion protection. In this case, metal-based smart–hybrid nanomaterials were widely used. The metal of smart–hybrid nanomaterials deposited on the metal surface; as a result, the cathodic hydrogen evolution was maximally stopped.

(vi) The energetic barriers were formed on the metal surface by the influence of smart–hybrid nanomaterials. The activation energy of corrosion ions is key in the high destruction and corrosion of materials. When the ions are more active, the metal destruction occurs. In comparison, the energetic barriers are mainly attributed to lessening the activation energy. As a result, the corrosive ions were deactivated.

12.2.3 Inorganic compound-based smart–hybrid nanomaterials in corrosion protection

Some self-healing smart–hybrid nanomaterialswere prepared from the inorganic nanocoatings. These types of nanomaterials are low-cost and more effective in corrosion protection. The metal surface was covered with these types of nanocoatings in chemical and electrochemical ways. The following metals were used in this self-healing of smart–hybrid nanomaterials: zinc-, nickel-, cerium-, potassium-, phosphate-, vanadium-, aluminum-, chromium-, zirconium- and molybdenum-rich layers [30, 31].

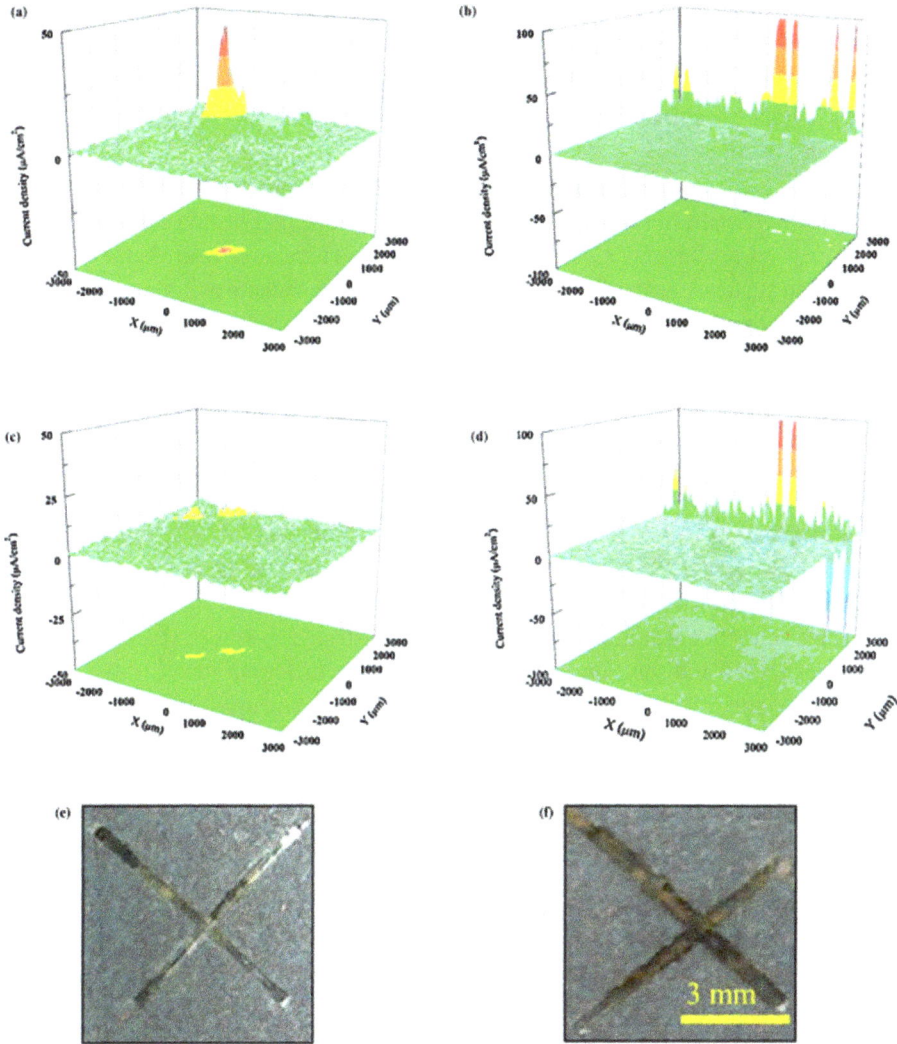

Figure 12.3: SVET results (120 min): (a) diethanolamine and (b) ethanolamine. SVET results (300 min): (c) diethanolamine and (d) ethanolamine. SEM results: (e) diethanolamine and (f) ethanolamine [29].

The basic task of metal-type smart–hybrid nanomaterials is that they raise polarization resistance and reduce the corrosion rate and corrosion potential. For example, the high polarization resistance is important in galvanized corrosion, in which the cold-rolled and hot-galvanized steel was protected by the metal-type smart–hybrid nanomaterials [32, 33].

In agriculture, automobile and other industries, phosphate-based smart–hybrid nanomaterials were widely used in metal corrosion protection. These types of materials were also performed in decoration, electrical insulation, wear reduction and metal-forming lubricants. The following basic components were used in the preparation of phosphate-based smart–hybrid nanomaterials: manganese or zinc salts, peroxides, chlorate and nitrite. In addition, silicon, permanganate, titanium salts and rare earth element phosphates were also used in the preparation of smart–hybrid nanomaterials [22, 34].

As observed, the manganese and zinc phosphate-based smart–hybrid nanomaterials were dominant in corrosion protection in the neural, acidic and saline solutions. The corrosion performance of these nanomaterials correlated with their structural factors and surface performances. The silica and zinc phosphate combinations can increase the corrosion resistance in the galvanic corrosion for steel-based metallic materials.

Zhang et al. [35] prepared lanthanum-based smart–hybrid nanomaterials for steel-galvanized corrosion. This nanomaterial is deposited on the metal surface, and lanthanum promotes corrosion cracks. In this research work, the nanomaterial growth on the metal surface was investigated. It was found that zinc influences the high growth rate of nanocoatings on the metal surface (Figure 12.4a). The next factor is the immersion time, and the growth of nanocoatings is rapid with the rise of immersion time. It reached a high growth rate after 1 h.

The electrochemical impedance results are illustrated in Figure 12.4b. The polarization resistance and solution resistance are the next factors in identification of corrosion efficiency. The influence of nanomaterials on the corrosion and inhibition mechanism was also investigated by the electrochemical impedance spectroscopy method. The obtained results confirmed that the lanthanum ions are mainly responsible for the high corrosion resistance and polarization resistance.

12.2.4 Organic compound-based smart–hybrid nanomaterials in the corrosion protection

The organic compounds and silica-based nanomaterials were mostly used in corrosion protection. These nanocoatings contain more halogen, alkoxy, hydroxyl, carboxyl and hydrolyzable functional groups. These types of materials are also named organosilane nanomaterials. The protective coatings are missed with organosilane to form the hybrid nanomaterials. The corrosion protection of these materials is very high because of their more rigid organosilane layer on the metal surface. The protective film was formed between the polymer coating and the surface of the metal. The organosilane was attached to the metal surface by the –O–Si– groups. The oxygen atoms covalently interact with the metal surface, and then the covalent bonds are formed between the metal substrate and organosilane. In comparison, the organic functional groups support the interactions and connections between organosilane and polymer-based coatings [12, 19].

The silicate–phosphate–molybdate is also widely performed in corrosion protection as the organic compound-based smart–hybrid nanomaterials. These types of

(a)

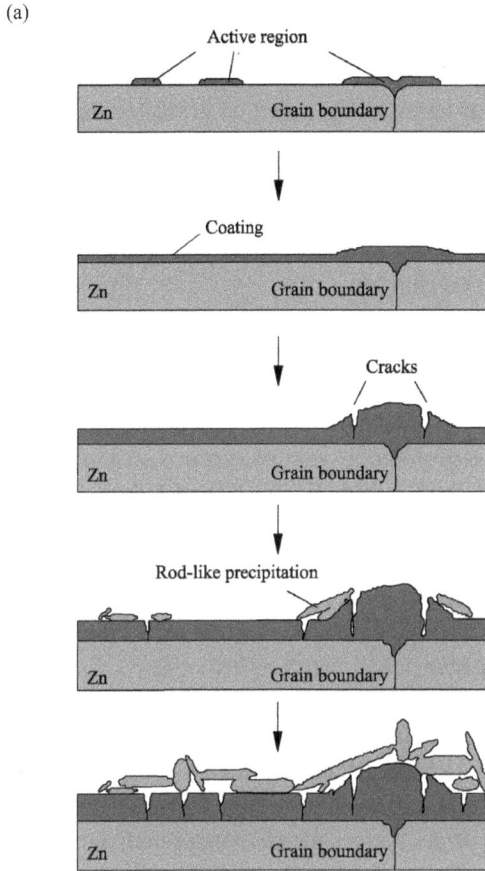

(b)

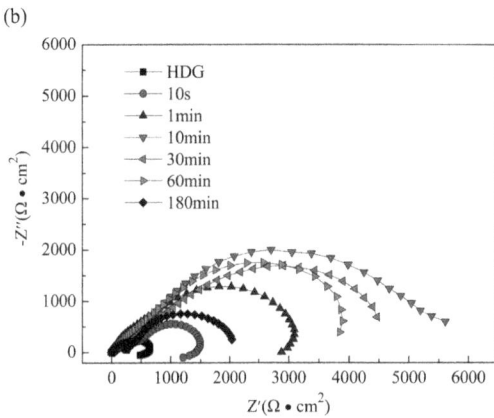

Figure 12.4: The protection mechanism of lanthanum-based smart–hybrid nanomaterials [35].

nanomaterials can protect steel metallic materials from corrosion in the sodium chloride environment. The outer interactions were supported by the Si–O bonds, while the Zn–P–Mo bonds are mainly responsible for the inner interactions [5, 11].

Figure 12.5 shows the preparation of nanocoating based on graphene oxide and tetraethoxysilane, and their corrosion protection mechanism is obtained in nanocoating [8]. This protective nanocoating can interact with the metal surface by the formation of Si–O–Me and C–O–Me bonds. This material is porous and has more pores. The silica polymers are attributed to the rise in the porosity of nanomaterials. The pore regions were illustrated in the corrosion mechanism.

The next smart feature of graphene oxide and tetraethoxysilane-based nanomaterials is the superhydrophobic surface. The metal surface was effectively insulated from the aquatic interactions by the superhydrophobic surface. The polarization resistance dramatically rose with the protective formation layer. The polarization properties, such as metal dissolution, hydrogen evolution, corrosion processes and anodic and cathodic electrochemical processes, were maximally blocked with the protective formation layer.

On the other hand, zinc ions or oxides were added to the smart nanocoatings. This is due to the antimicrobial protection. The microbial or bacterial corrosion of metals is also a problem. For example, the aquatic phase of crude oil is more microbial. Zinc is promoted to corrosion resistance in antimicrobial activities.

In modern industry, smart–hybrid nanomaterials are widely employed. UV radiation is also a harmful factor for the protective layer on the metal surface. Smart nanomaterials have a self-protect mechanism against UV radiation. The old nanocoatings are more sensible for UV radiation; as a result, their protective layer was easily destroyed by UV radiation. Nevertheless, smart nanocoatings are more efficient in UV radiation, and their lifetime is very long [18, 27].

12.3 Conclusion

Smart-hybrid nanoparticles come in three varieties: superhydrophobic, self-healing, and self-cleaning. The silica, epoxy, polymer, and carbon-based components were used to create the smart-hybrid nanomaterials. These smarthybrid nanomaterials have a number of beneficial qualities. For instance, nanoparticles based on polymers, silica, and epoxy have remarkable mechanical and thermal stability.

From the inorganic nanocoatings, certain self-healing smart-hybrid nanomaterials were created. These particular nanoparticles are less expensive and better at preventing corrosion. These kinds of nanocoatings were applied to the metal surface using chemical and electrochemical processes. The zinc, nickel, cerium, potassium, phosphate, vanadium, aluminum, chromium, zirconium, and molybdenum-rich layers were utilised in this self-healing of smart-hybrid nanomaterials.

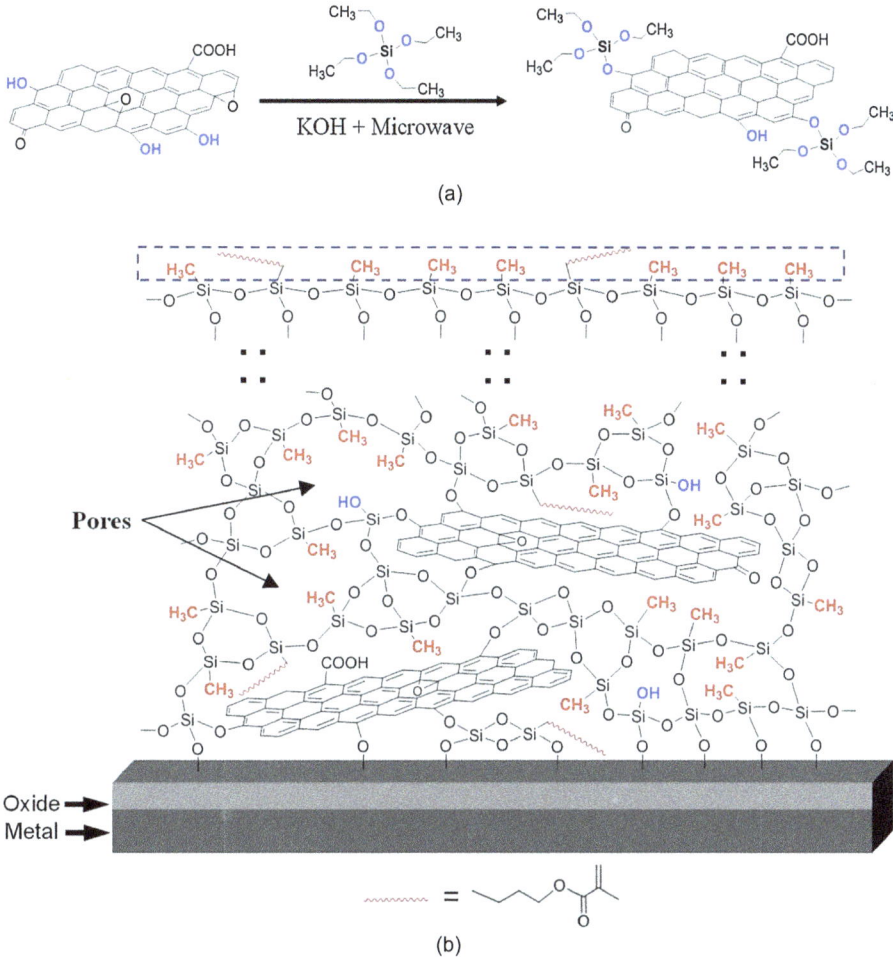

Figure 12.5: (a) Preparation of nanocoating based on graphene oxide and tetraethoxysilane and (b) corrosion protection mechanism of obtained nanocoating [36].

Nanocoatings were used to protect the metal materials. Nanomaterials that are intelligent and hybrid have become popular for corrosion prevention. This is because smart-hybrid nanomaterials, which form a barrier oxide layer on the metal surface, have some advantages in terms of protecting metals. This film has a thickness of a few nanometers or less. These characteristics mostly account for how well nanocoatings resist corrosion. The most popular nanomaterials for corrosion protection were those made of silica and organic compounds. More halogen, alkoxy, hydroxyl, carboxyl, and hydrolyzable functional groups are present in these nanocoatings. Organosilane nanoparticles is another name for these substances. Organosilane is omitted from the protective coatings to create hybrid nanomaterials.

References

[1]   Yan, D., et al., Dual-functional graphene oxide-based nanomaterial for enhancing the passive and active corrosion protection of epoxy coating, Composites Part B: Engineering, 2021, 222, 109075.

[2]   Tavandashti, N.P., Almas, S.M., Esmaeilzadeh, E., Corrosion protection performance of epoxy coating containing alumina/PANI nanoparticles doped with cerium nitrate inhibitor on Al-2024 substrates, Progress in Organic Coatings, 2021, 152, 106133.

[3]   Zheludkevich, M.L., Salvado, I.M., Ferreira, M.G.S., Sol–gel coatings for corrosion protection of metals, Journal of Materials Chemistry, 2005, 15(48), 5099–5111.

[4]   Wieszczycka, K., Staszak, K., Woźniak-Budych, M.J., Litowczenko, J., Maciejewska, B.M., Jurga, S., Surface functionalization–The way for advanced applications of smart materials, Coordination Chemistry Reviews, 2021, 436, 213846.

[5]   Yeganeh, M., Nguyen, T.A., Rajendran, S., Kakooei, S., Li, Y., Corrosion protection at the nanoscale: An introduction, in Corrosion protection at the nanoscale, Elsevier, 2020, 3–7.

[6]   Sharafudeen, R., Chapter 10 – Smart hybrid coatings for corrosion protection applications, in Makhlouf, A.S.H., Abu-Thabit, N.Y. (Eds.), Advances in Smart Coatings and Thin Films for Future Industrial and Biomedical Engineering Applications, Elsevier, 2020, 289–306.

[7]   Wen, J., Lei, J., Chen, J., Gou, J., Li, Y., Li, L., An intelligent coating based on pH-sensitive hybrid hydrogel for corrosion protection of mild steel, Chemical Engineering Journal, 2020, 392, 123742.

[8]   Santos, L.R.L., Marino, C.E.B., Riegel-Vidotti, I.C., Silica/chitosan hybrid particles for smart release of the corrosion inhibitor benzotriazole, European Polymer Journal, 2019, 115, 86–98.

[9]   Nazeer, A.A., Madkour, M., Potential use of smart coatings for corrosion protection of metals and alloys: A review, Journal of Molecular Liquids, 2018, 253, 11–22.

[10] Nazari, M.H., et al., Nanocomposite organic coatings for corrosion protection of metals: A review of recent advances, Progress in Organic Coatings, 2022, 122, 106573.

[11] Ulaeto, S.B., Pancrecious, J.K., Rajan, T.P.D., Pai, B.C., Smart coatings, in Noble Metal-Metal Oxide Hybrid Nanoparticles, Elsevier, 2019, 341–372.

[12] Makhlouf, A.S.H., Abu-Thabit, N.Y., Advances in smart coatings and thin films for future industrial and biomedical engineering applications, Elsevier, 2019.

[13] Figueira, R.B., Sousa, R., Silva, C.J.R., Multifunctional and smart organic–inorganic hybrid sol–gel coatings for corrosion protection applications, Advances in Smart Coatings and Thin Films for Future Industrial and Biomedical Engineering Applications, 2020, 57–97.

[14] Khan, A., Hassanein, A., Habib, S., Nawaz, M., Shakoor, R.A., Kahraman, R., Hybrid halloysite nanotubes as smart carriers for corrosion protection, ACS Applied Materials & Interfaces, 2020, 12 (33), 37571–37584.

[15] Montemor, M.F., Functional and smart coatings for corrosion protection: A review of recent advances, Surface and Coatings Technology, 2014, 258, 17–37.

[16] Jothi, K.J., et al., Fabrications of hybrid polyurethane-Pd doped ZrO2 smart carriers for self-healing high corrosion protective coatings, Environmental Research, 2022, 211, 113095.

[17] Zheludkevich, M.L., Tedim, J., Ferreira, M.G.S., "Smart" coatings for active corrosion protection based on multi-functional micro and nanocontainers, Electrochimica Acta, 2012, 82, 314–323.

[18] Jeong, H.H., Alarcón-Correa, M., Mark, A.G., Son, K., Lee, T.C., Fischer, P., Corrosion-protected hybrid nanoparticles, Advanced Science, 2017, 4(12), 1700234.

[19] Ashrafi-Shahri, S.M., Ravari, F., Seifzadeh, D., Smart organic/inorganic sol-gel nanocomposite containing functionalized mesoporous silica for corrosion protection, Progress in Organic Coatings, 2019, 133, 44–54.

[20] Sharafudeen, R., Smart hybrid coatings for corrosion protection applications, in Advances in Smart Coatings and Thin Films for Future Industrial and Biomedical Engineering Applications, Elsevier, 2020, 289–306.

[21]   Kaseem, M., Ko, Y.G., A novel hybrid composite composed of albumin, WO3, and LDHs film for smart corrosion protection of Mg alloy, Composites Part B: Engineering, 2021, 204, 108490.

[22]   Figueira, R.B., Fontinha, I.R., Silva, C.J.R., Pereira, E.V., Hybrid sol-gel coatings: Smart and green materials for corrosion mitigation, Coatings, 2012, 6(1), 12.

[23]   Sanchez, C., Shea, K.J., Kitagawa, S., Hybrid materials, Chemical Society Review, 2011, 40, 453–1152.

[24]   S Tankiewicz, A., 14 – Self-healing nanocoatings for protection against steel corrosion, in Pacheco-Torgal, F., Diamanti, M.V., Nazari, A., Granqvist, C.G., Pruna, A., Amirkhanian, S. (Eds.), Nanotechnology in Eco-efficient Construction, Second Edition, Woodhead Publishing, 2019, 303–335.

[25]   Zea, C., Alcántara, J., Barranco-García, R., Morcillo, M., De la Fuente, D., Synthesis and characterization of hollow mesoporous silica nanoparticles for smart corrosion protection, Nanomaterials, 2018, 8(7), 478.

[26]   Samiee, R., Ramezanzadeh, B., Mahdavian, M., Alibakhshi, E., Assessment of the smart self-healing corrosion protection properties of a water-base hybrid organo-silane film combined with non-toxic organic/inorganic environmentally friendly corrosion inhibitors on mild steel, Journal of Cleaner Production, 2019, 220, 340–356.

[27]   Gonçalves, J.L.M., Crucho, C.I.C., Alves, S.P.C., Baleizão, C., Farinha, J.P.S., Hybrid mesoporous nanoparticles for pH-actuated controlled release, Nanomaterials, 2019, 9(3), 483.

[28]   Chen, Z., Yang, W., Chen, Y., Yin, X., Liu, Y., Smart coatings embedded with polydopamine-decorated layer-by-layer assembled SnO2 nanocontainers for the corrosion protection of 304 stainless steels, Journal of Colloid and Interface Science, 2020, 579, 741–753.

[29]   Choi, H., Kim, K.Y., Park, J.M., Encapsulation of aliphatic amines into nanoparticles for self-healing corrosion protection of steel sheets, Progress in Organic Coatings, 2013, 76(10), 1312–1324.

[30]   Li, H., Qiang, Y., Zhao, W., Zhang, S., 2-Mercaptobenzimidazole-inbuilt metal-organic-frameworks modified graphene oxide towards intelligent and excellent anti-corrosion coating, Corrosion Science, 2021, 191, 109715.

[31]   Shchukina, E., Shchukin, D., Grigoriev, D., Effect of inhibitor-loaded halloysites and mesoporous silica nanocontainers on corrosion protection of powder coatings, Progress in Organic Coatings, 2017, 102, 60–65.

[32]   Chen, T., Fu, J., An intelligent anticorrosion coating based on pH-responsive supramolecular nanocontainers, Nanotechnology, 2012, 23(50), 505705.

[33]   Khasim, S., Pasha, A., Enhanced corrosion protection of A-36 steel using epoxy-reinforced CSA-doped polyaniline-SnO$_2$ nanocomposite smart coatings, Journal of Bio- and Tribo-Corrosion, 2021, 7, 1–11.

[34]   Li, G.L., Zheng, Z., Möhwald, H., Shchukin, D.G., Silica/polymer double-walled hybrid nanotubes: Synthesis and application as stimuli-responsive nanocontainers in self-healing coatings, ACS Nano, 2013, 7(3), 2470–2478.

[35]   Zhang, S.-H., Kong, G., Lu, J.-T., Che, C.-S., Liu, L.-Y., Growth behavior of lanthanum conversion coating on hot-dip galvanized steel, Surface and Coatings Technology, 2014, 259, 654–659.

[36]   Li, J., Cui, J., Yang, J., Ma, Y., Qiu, H., Yang, J., Silanized graphene oxide reinforced organofunctional silane composite coatings for corrosion protection, Progress in Organic Coatings, 2012, 99, 443–451.

Index

https://doi.org/10.1515/9783111071756-013

www.ingramcontent.com/pod-product-compliance
Lightning Source LLC
Chambersburg PA
CBHW061420210326
41598CB00035B/6284

* 9 7 8 3 1 1 1 0 7 0 0 9 4 *